AF872278

PRACTICAL AI STRATEGIES FOR BUSINESS GROWTH

DESIGNATED AI

MARCIN LADOWSKI, TARIK SAFOUAN, JENS MÜLLER

Copyright © 2026 by **Marcin Ladowski, Tarik Safouan, and Jens Müller.**
Published by **Designated AI**
All rights reserved.

No part of this publication may be reproduced, distributed, transmitted, stored in a retrieval system, or used in any form or by any means, electronic, mechanical, photocopying, recording, scanning, or otherwise, without the prior written permission of the copyright holders, except for brief quotations used in reviews, criticism, or other uses permitted by applicable copyright law.

This book is intended for educational and informational purposes only. The information, strategies, examples, frameworks, and recommendations contained in this publication are provided in good faith and are not intended to constitute legal, financial, investment, technical, or other professional advice. Readers are responsible for evaluating the suitability of any information or strategy for their own circumstances and organizations.

Artificial intelligence technologies, tools, platforms, and capabilities evolve rapidly. References to technologies, products, services, companies, or platforms are provided for informational purposes and do not necessarily constitute an endorsement or recommendation. Readers should independently verify current capabilities, terms, privacy practices, security requirements, and applicable regulations before implementing any AI system.

Every effort has been made to present information accurately at the time of publication. However, the authors and publisher make no warranties or representations regarding the completeness, accuracy, reliability, or continued applicability of the information contained in this book.

Some examples and business situations in this book are illustrative and may be adapted or fictionalized for educational purposes. Any resemblance to actual persons, organizations, or events beyond those expressly identified is coincidental.

Paperback ISBN: 979-8-9976681-0-5
Hardcover ISBN: 979-8-9976681-1-2

For additional resources, worksheets, prompts, checklists, and companion materials, visit:

https://designated.ai/books/practical-strategy

Paperback ISBN: [illegible]
Hardcover ISBN: [illegible]

Copyright © 2026 by **Marcin Ladowski, Tarik Safouan, and Jens Müller.**
Published by **Designated AI**
All rights reserved.

No part of this publication may be reproduced, distributed, transmitted, stored in a retrieval system, or used in any form or by any means, electronic, mechanical, photocopying, recording, scanning, or otherwise, without the prior written permission of the copyright holders, except for brief quotations used in reviews, criticism, or other uses permitted by applicable copyright law.

This book is intended for educational and informational purposes only. The information, strategies, examples, frameworks, and recommendations contained in this publication are provided in good faith and are not intended to constitute legal, financial, investment, technical, or other professional advice. Readers are responsible for evaluating the suitability of any information or strategy for their own circumstances and organizations.

Artificial intelligence technologies, tools, platforms, and capabilities evolve rapidly. References to technologies, products, services, companies, or platforms are provided for informational purposes and do not necessarily constitute an endorsement or recommendation. Readers should independently verify current capabilities, terms, privacy practices, security requirements, and applicable regulations before implementing any AI system.

Every effort has been made to present information accurately at the time of publication. However, the authors and publisher make no warranties or representations regarding the completeness, accuracy, reliability, or continued applicability of the information contained in this book.

Some examples and business situations in this book are illustrative and may be adapted or fictionalized for educational purposes. Any resemblance to actual persons, organizations, or events beyond those expressly identified is coincidental.

Paperback ISBN: 979-8-9976681-0-5
Hardcover ISBN: 979-8-9976681-1-2

For additional resources, worksheets, prompts, checklists, and companion materials, visit:

https://designated.ai/books/practical-strategy

Acknowledgments

Marcin:

To my wife Cintia, I am grateful for the trust and support in pursuing my CXO program at MIT while writing this book. You inspired me to write it for professionals like you, who see the opportunities AI can bring to their business.

Lucia and Mateo, watching you grow up curious and brave gives me the energy to do hard things, and reminds me every day why they're worth doing. #HomeTeam

Jens and Tarik, I appreciate your trust in taking this on together and your endurance through the process. Jens, your discipline and care shaped every part of this. Tarik, your sharp thinking made the work better every time we talked.

Irma, I am grateful for the way you kept three authors moving in the same direction and made it look easy. It wasn't. This book exists because you made the collaboration work.

To my MIT professors and mentors: Dr. David Niño, Dr. George Westerman, Dr. Abel Sanchez, Blade Kotelly, and many others. Fifteen months of your teaching reshaped how I think about leadership and technology.

Scott and Sam, thanks for the early reads and honest feedback.

Tarik:

I want to begin by thanking my family and friends for their support, patience, and encouragement throughout this journey. A very special thank you to Marcin, who invited me to work on this book and turned what began as an idea into a fun and rewarding two-year journey of writing, debating, learning, and discovering together.

I am deeply grateful to Dr. Bogdan Daraban, Dean of the Andreas School of Business, for giving me the opportunity and trust to put AI strategy into practice through the AI Center and to develop the MBA specialization in AI Strategy. I also want to thank Dr. Lewis and Dr. Katsiaryna Matusevich, as well as the many students at Barry University who, through conversations, questions, projects, experiments, and sometimes disagreements, contributed indirectly to many of the ideas and lessons reflected in this book.

My sincere thanks also go to Irma, our outstanding project manager, whose dedication, structure, and care were essential in bringing this book together, and to my co-author Jens, who has been a constant source of inspiration through everything he is building in technology.

My thanks also extend to the Branches community for becoming part of our journey of learning, experimenting, and exchanging ideas around AI, with a special thank you to Steve Arman for his support and engagement.

Finally, I want to remember Dr. Daniel Rosenberg, who introduced me to academia and to teaching. That introduction opened a path I could not have fully imagined at the time and ultimately became an important part of the journey that led to this book. I remain deeply grateful for his influence and guidance. May he rest in peace.

Jens:

Writing a book is never the work of one person alone. Although only a few names appear on the cover, every page reflects the encouragement, support, and contributions of many people who have accompanied this journey. I would therefore like to express my sincere gratitude to those who made this book possible.

First and foremost, I would like to thank my family. Writing this book required countless evenings, weekends, and moments that would otherwise have been spent together. Their patience, understanding, encouragement, and unwavering support gave me the time and freedom to devote myself fully to this project. Their belief in me never wavered, even when the work seemed endless, and for that I am deeply grateful.

My heartfelt thanks also go to Marcin, who invited me to become part of this book as a co-author. His trust, vision, and enthusiasm created the opportunity for me to contribute to this project from the very beginning. Throughout the writing process, our many discussions challenged ideas, inspired new perspectives, and ultimately helped shape many of the concepts presented in these chapters. I am equally grateful to Tarik for the numerous insightful conversations we shared. The exchange of ideas, experiences, and viewpoints enriched this book far beyond what could have been achieved individually.

A project of this scope also depends on excellent coordination, and for that I owe my sincere appreciation to Irma. As our project manager, she guided this project with remarkable dedication, professionalism, and attention to detail. Her organizational skills, persistence, and ability to keep everyone aligned ensured that the project continued to move forward, even during its most demanding phases. Her commitment played a significant role in bringing this book to completion.

Finally, I would like to thank everyone who has contributed, directly or indirectly, to my professional journey over the years. Colleagues, mentors, customers, research partners, and friends have continually challenged me to think differently, ask better questions, and remain curious about new technologies. Many of the ideas presented in this book were inspired by real-world discussions, practical challenges, and shared experiences across industry and research. I hope that the knowledge collected in these pages will help others approach Artificial Intelligence with confidence, curiosity, and a clear focus on creating meaningful value.

Our sincere thanks also to our test readers, their thoughtful and constructive feedback helped us refine and improve the content significantly

Kevin Larkin, M.Sc. in Materials Science and Engineering
PhD Candidate at Carnegie Mellon University, Computational Mechanics from the Civil and Environmental Engineering Department Pittsburgh, Pennsylvania, USA

Priyanka Ghosh, M.Sc. in Electrical and Electronics
PhD Candidate at Helmut Schmidt University, Department of Power Electronics, Hamburg, Germany

Marcel Eckert, Dr.-Ing. Academic Director
Helmut Schmidt University, Department of Computer Engineering Hamburg, Germany

Miriam Prohl, B.A. in Socioeconomics
Facilities Team Leader, DB Deutsche Bahn Hamburg, Germany

Oliver Schröder, Dipl.-Ök. in Economics
Senior Specialist Resource & Project Management, DB Engineering & Consulting Berlin, Germany

Christine Larkin
Senior Documentation Manager, Jellyfish
Boston, Massachusetts, USA

Lars Lehmann, Dipl.-Ing. in Mechanical Engineering
Engineering Manager responsible for New Developments across All Programs at Airbus (e.g. A380, A350, R&T, In-Service, LR, SA) Hamburg, Germany

Ralf Nagel, Dipl.-Phys. in Physics
CEO at DYNAInstruments GmbH
Hamburg, Germany

Table of Contents

Preface

I've always had a thing for quality. My friends and family like to joke that I can't leave well enough alone, that I'll always make "just one more tweak." As a kid, I grew up hearing sayings like "Do it well, or don't do it at all" and "How you do anything is how you do everything." Somewhere along the way, those lessons stuck. It's a blessing most of the time, though occasionally a curse if you ask the people around me.

That's why, when my friend Tarik called one afternoon with an unexpected invitation, my first reaction was hesitation. Tarik, one of the co-authors of this book, was speaking at a Small Business Summit on AI. A slot had opened up at the last minute, and he asked if I could step in.

The audience was exactly the people I care most about supporting, leaders of small and mid-sized businesses. The topic couldn't have been more timely: the opportunities and risks of adopting AI. Best of all, it was local, right here in South Florida.

There was only one problem. I had one week to prepare.

English is my second language, and while I've done plenty of public speaking, I usually need time to prepare. Normally, I'd spend weeks sketching the outline, rehearsing, polishing slides, and finding the rhythm of the talk. A week didn't feel like enough.

But this time was different. This time, I had AI at my side.

I used a large language model to generate ideas and sharpen my message. An AI presentation tool gave me a starter deck I could refine and build on. An AI coaching app flagged issues with my pacing and delivery and gave me real-time feedback as I rehearsed on camera.

Because of those tools, I said yes. The session went well, and the conversations afterward told me the topic had struck a nerve.

That week changed how I view AI. I realized how many people, business owners, managers, and even entire companies, say no to opportunities simply because they lack confidence or don't know where to start. The world is moving faster than ever, but we don't have to be left behind.

Most small business leaders struggle with the same few constraints. Money, time, and access to knowledge. You often hear that time is money, but the real formula is broader. If you have time and expertise, money usually follows. The more you know, and the better you use your time, the stronger your business becomes.

New technology can give you a richer, more impactful working life. Automate the redundant and tedious tasks, and you free up time to connect more deeply with your clients, focus on what truly matters, and build amazing things.

That's exactly what AI did for me. It let me prepare in a fraction of the time and at a fraction of the cost. It gave me leverage, the ability to do in one week what would normally take four. I seized an opportunity I'd otherwise have let pass me by.

As exciting as the technology is, people still matter most. AI can speed you up and make you more productive, but it'll never replace a great team. If you want to go far and have your work truly matter, you need others by your side.

Team

This book was written as a collaborative effort. To make it possible, I brought together colleagues whom I deeply respect, each with expertise in this topic but from different industries and countries, to bring diverse perspectives to this work. That is why the voice may shift as you read, and we've chosen to let it. Each of us writes best about what we know best. What unites us is a shared purpose: helping small and mid-sized businesses harness AI in a way that is practical, sustainable, and built to last.

Marcin

Before I introduce my co-authors, a few words about myself. I've spent my career as an operator. I founded and ran my own company and later served as COO of a family-owned business, where I helped it professionalize and scale. These days, I advise small and mid-sized businesses on strategy, automation, and AI adoption. Building products has always been my first love, and I'm at it again now, working on an AI-powered platform in the talent management space. I've also gone back to school, first through MIT's Applied Data Science program, where I met Jens, and more recently through MIT's Chief Product Officer program, because this field moves fast and I'd rather learn alongside it than pretend I've finished. I've gotten plenty of things wrong along the way, chased a few shiny objects myself, and learned most of what's in this book the slow way. That experience, more than any credential, is what I hope we can offer you here.

Tarik

I first met Tarik a few years ago, and from the first conversation, it was clear he brought a rare blend of strategic thinking, practical experience, and real curiosity. His background stood out immediately. A graduate of Harvard University, he learned early how to operate in fast and demanding environments. He has worked across e-commerce, media, aviation, agency work, and consulting, with organizations including Gannett, Formula 1, and the Miami Dolphins, and contributed to projects with La Liga, the Premier League, and the Miami Open. One of the first things we connected on was our shared love of tennis.

Tarik made a strong impression right away. He is the kind of person you hope to cross paths with again. He understands data, culture, operations, and people at a deep level, and he brings those strengths together in a way that is both grounded and forward-thinking. Today, he leads the AI Center at Barry University, shaping the future of innovation and technology education. I respect him enormously, and I'm grateful our careers aligned in a way that lets us build this project together.

Jens

I met Jens a few years ago during one of the Applied Data Science programs at MIT, and from the first conversation, I knew he was someone special. He is an engineer in the truest sense of the word, the kind who can design and build complex systems from the ground up. His work in the automotive industry, his leadership at TBench.Solutions in Germany, and his ability to bridge mechanical engineering with advanced AI concepts give him a perspective that is incredibly rare. Jens holds multiple patents, but what stands out to me even more is his mindset. He brings technical depth and a genuine optimism about the role technology can play in improving people's lives.

He understands AI at a foundational level and how it behaves in real-world systems, manufacturing floors, and innovation projects. Every time we speak, I'm reminded why I value him so much as a collaborator. Jens sees the world through possibility, and having him on this project brings a level of rigor and humanity that strengthens everything we're building together.

Introduction

You've heard that AI is changing everything, and you've probably wondered whether any of it is built for a company your size. It is. This book is written for you.

What happens when a strategist, an operator, and an engineer write a book together? Every chapter takes longer than planned. We see business through different lenses, shaped by different countries and different careers, and those lenses are not neutral. Neither are yours.

We each wrote the chapters we know best, so the voice changes as you go. You'll notice the handoff. We left it in.

The mix is valuable because the perspective you are missing is the one most likely to cost you. That difference shaped every decision we made while writing this book.

Here's the part we want you to hear clearly. You are not late. The businesses pulling ahead right now didn't start years before you. That gap is still wide open. Capability is almost never what stops a company. Confusion, hype, and the sheer volume of options stop it, long before anyone runs into a real limit of what the technology can do.

AI has stopped being optional. It has become a strategic lever, the kind that changes what a small team can carry and how fast a good idea reaches a customer. Pulling that lever takes clarity more than it takes budget or engineers.

That's why we wrote this book. It gives you a way to see the whole landscape, a structure for deciding what deserves your attention, and a first move you can make this week. It shows how strong businesses have always been built on systems, people, and process, with AI as the newest layer on a foundation that still matters. It arms you against shiny-object syndrome, so your energy follows your strategy rather than the loudest trend of the month. It's built as a playbook, meant to be flipped through, tested, marked up, and returned to when the next decision lands on your desk.

AI is a fascinating subject, and it can also feel abstract. To keep it grounded, we've woven short stories, fables, and real business situations through the chapters. Each one takes an idea you're about to meet, or have just met, and gives it a shape you can picture and carry with you. When you come across one, treat it as part of the explanation. The story and the plain-language breakdown around it are doing the same work from different angles, and you'll get the most out of a chapter by reading both.

What you won't find here is a long march through technical machinery or a catalog of tools that would be dated before you finish the book. Plenty of books already cover that ground. Our focus stays on guidance you can act on, situations you'll recognize from your own business, and examples you can put to use immediately.

The book also reaches past its own pages. We maintain a companion resource page online and keep adding to it as the technology moves.

Whenever you see a code like the one beside this paragraph, point your phone camera at it, and you'll land on the material tied to that chapter, whether that's a worksheet, a checklist, a set of prompts, or a diagram you can put in front of your team on Monday.

Scan for this chapter's resources

You can also reach everything directly at: https://designated.ai/books/practical-strategy

Our aim is that you close this book equipped and confident, curious enough to try something new, and clear enough on the steps to actually do it.

Those goals guided the principles we followed while writing this book.

To meet you where you are, in this book, we commit to:

Clarity over jargon.

We'll keep things simple, practical, and free of fluff or endless tool lists.

Real stories, real results.

You'll see how AI is solving everyday problems for small and mid-sized businesses.

Action you can take.

We'll share the 4A AI Strategy Framework and step-by-step approaches you can apply immediately, no matter your background.

Honesty first.

We'll be upfront about what works, what doesn't, and what's better left for later.

Confidence to say yes.

Most of all, we'll help you embrace opportunities that once felt intimidating or out of reach.

This manual helps you and your team adopt AI the right way. Less friction, lower cost, stronger teams, room to scale. The future of business belongs to leaders who use AI consistently and with clarity. That future is within your reach.

AI is here to stay. The sooner it becomes your tool, the sooner you'll become the leader in your field.

So turn the page, and let's get started.

This is your chance to say YES!

Part 1

FOUNDATION

Chapter 1.1

What AI Really Is

When it comes to artificial intelligence, most people fall into one of two camps. Some see it as an abundant resource, a force that will help us work smarter, grow faster, and thrive. Others find it unsettling, even threatening, and would rather keep it at arm's length. No matter which camp you find yourself in, the same thing tends to be missing: a clear, grounded sense of what AI actually is and how it works. That is what this chapter is for. We will keep it practical and free of jargon, and we will begin somewhere familiar, with an ordinary day.

Imagine you are traveling abroad and your credit card suddenly gets blocked at a restaurant. Before the panic sets in, your phone vibrates with a message: "Did you just attempt an $850 purchase at a boutique in Paris, France?"

You blink. You are in London for a conference. Paris is definitely not on your itinerary. You type "NO," and just like that, the fraudulent transaction is stopped, your card is secured, and your sense of security is restored. You go back to your coffee, grateful for the quick catch.

A few days later, you breeze through airport security. You pause, look at the camera, and in a fraction of a second, the system checks whether your face matches the one on your passport. The green light flashes. You walk through without a line and without stress, knowing a system made that decision in an instant so officers could focus on what actually needs their attention.

Back home, you find yourself at your doctor's office after a minor health scare. The X-ray runs. What you do not see is that before a radiologist opens it, a system has already sorted the day's images by urgency, weighing each against patterns learned from prior scans. The urgent ones move to the front of the queue. Yours was not one of them, and the report comes back clear [1].

All of this happened quietly, without any grand announcement or big-screen moment. You have already interacted with AI many times, often without realizing it. It is there in your bank, at the airport, in your doctor's office, and across the small moments that make up an ordinary day.

Breaking Down the Buzzword

The phrase Artificial Intelligence can sound intimidating. It tends to call up images of complex machines and advanced mathematics, the kind of thing that belongs to specialists. But broken into its two plain words, it becomes far less mysterious.

Artificial simply means human-made. These systems do not appear on their own. We build them. And intelligence, in this context, as it is now it means the ability to learn from experience and use what was learned to make a decision. Put the two together and artificial intelligence (AI) is really just a human-made technology that learns from examples and uses that learning to make predictions or decisions.

There is a simpler way to think about what AI is really doing. At its heart, it works by probability, what specialists call probabilistic computation. It studies patterns in past data, then calculates the most likely outcome in a new situation. That word, likely, matters. AI does not know the future. It predicts it, the same way an experienced manager reads a quarter before the numbers are final.

The Model in Your Head

To make those predictions, an AI system builds a model. A model is simply a trained representation of patterns from past data, captured so it can be reused for a specific purpose. That sounds abstract until you realize you have built plenty of them yourself.

Think about your first drive to a new office. That first morning feels uncertain. You rely on turn-by-turn directions, checking every sign and second-guessing every exit, because all of it is new. But after a few weeks something changes. You stop reaching for the directions. You know the landmarks, the lane that backs up, the light that always seems to catch you. You can make the drive on instinct, and you can even tell a colleague how long it will take depending on when they leave.

That instinct is a model. It does not store a recording of every single drive. It pulls the patterns out of all those trips and turns them into a set of rough rules you can apply the next time. An AI model works the same way. It learns rules from data, then uses them to make quick, practical predictions when something new comes along.

The AI That's Been Hiding in Plain Sight

This pattern, learning from data and using it to predict, is not new. AI has been working quietly inside entire industries for decades, long before anyone was talking about it at dinner. It also comes in a few distinct forms, and you have already met most of them without knowing their names.

The foundation underneath nearly all of it is machine learning, the ability to improve at a task by learning from examples [2]. It is what banks have used to catch fraud since the 1990s, flagging the one odd transaction in a sea of ordinary ones [3]. It is also the engine that lets a business forecast next quarter's demand by studying years of sales history.

Built on top of that foundation are a few forms you will meet often. Computer vision gives machines the ability to interpret images, which is what lets a self-driving car tell a pedestrian from a roadside post and helps a doctor read a scan by estimating whether a shadow is worth a closer look.

Natural language processing handles written words, part of what keeps spam out of your inbox, and the foundation of the conversational tools we will come to shortly. Its close partner, speech recognition, does the same for the spoken word, powering the voice assistants that take dictation and the systems that now answer and transcribe business calls. And recommender systems are the quiet engines behind the products, shows, and articles suggested to you every day, each one reading patterns in what people did before.

For most of that history, though, AI was out of reach for the average business. AI did not exist on its own. Putting it to work meant standing at the crossroads of several demanding fields at once, computer science, mathematics and statistics, software development, research, and a deep understanding of the business problem itself. That is the world of the data scientist, the specialist trained to pull all of these threads together into something that works as depicted in Figure 1.1.

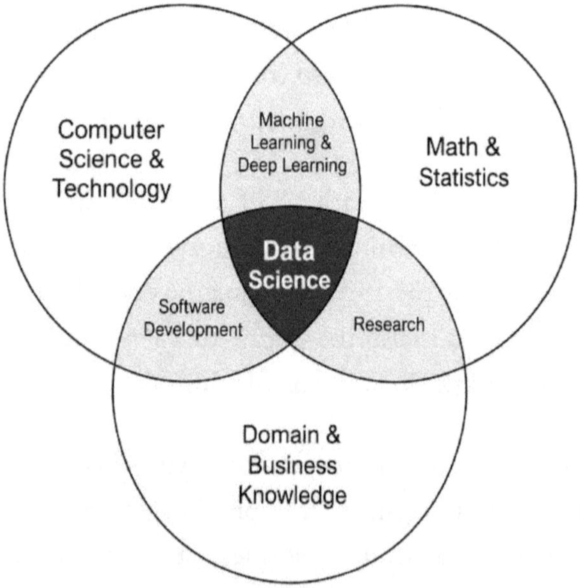

Figure 1.1 What it used to take. Using AI once meant standing at the center of many specialized fields, the world of the data scientist.

For a large, well-funded company, hiring that kind of talent was possible. For most small and medium businesses, it was not. The specialists were scarce and expensive, and building a system from scratch took months. So, for everyone outside the largest firms, AI stayed locked behind a door marked 'too costly' and 'too complicated' [4].

Why AI Suddenly Feels New

The version of AI most people first noticed, the kind you can hold a conversation with, is genuinely recent. But the technology underneath it is not. What changed was not the invention of AI, but the way we reach for it.

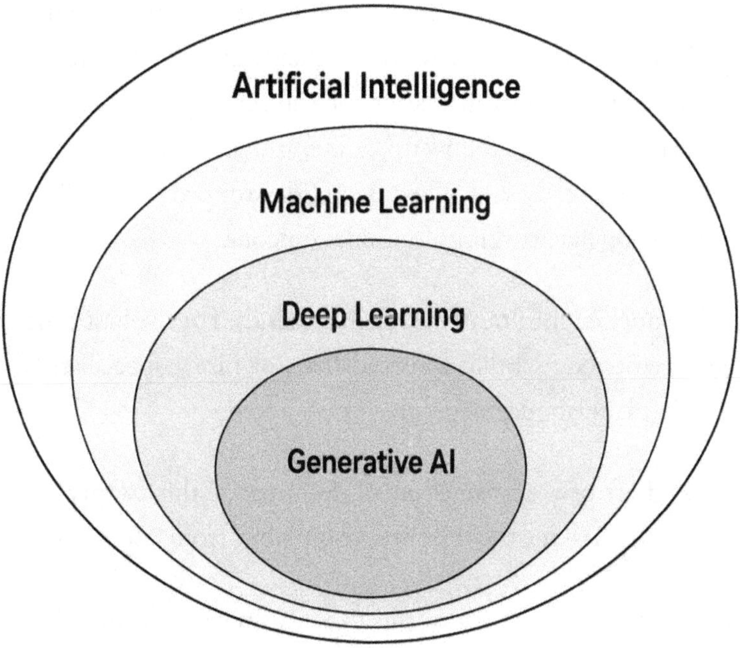

Figure 1.2 The overview of how everything fits into AI

Think about how we found information before. For twenty years, the search engine was the front door to human knowledge. You typed in a few keywords and got back a list of links. Useful, but the real work was still yours: opening the pages, comparing what they said, and piecing together your own answer. The search engine pointed you toward knowledge. It did not assemble it for you.

What it is now. The expertise moved inside the tools. As depicted in Figure 1.2, Artificial intelligence is a broad field, machine learning learns from data, deep learning is a more powerful form of it, and generative AI, the kind you can talk to, sits within.

Large language models, which are a subset of Deep Learning, changed that exchange. Instead of a list of places to look, they bring the relevant information directly to you, drafted into an answer you can actually use. The conversation also has a memory. You can follow up, refine, and push back, and it keeps track of what you have already said. The interaction stopped feeling like searching and started feeling like working alongside someone.

This is what puts AI into nearly everyone's hands. For the first time, you did not need to write code, build a spreadsheet, or hire a specialist. You could simply ask, in plain language, for what you wanted.

Write a first draft of a customer email. Summarize this twenty-page report into three points. Tell me the top three complaints from last month's reviews.

The sentence replaced the spreadsheet. That single shift turned AI from a capability locked inside large companies into something a manager, owner, or team lead could pick up and use directly. It stopped being software you had to operate and become more like a capable assistant you could talk to. This is a powerful starting point that opened the door to even more advanced, proactive collaboration.

Generative AI is an extension of the same idea. Instead of only predicting whether a transaction is fraudulent or where sales are heading, it predicts the most likely next word in a sentence, or the most likely image or piece of audio to match what was asked for [5]. String enough of those predictions together and it produces something new, a paragraph, a picture, a summary. Underneath the apparent creativity, it is still doing what AI has always done, recognizing patterns and playing the odds, now pointed at making things rather than only sorting them.

Not Magic, Just Method

The first time many leaders use a modern AI tool, it can feel almost like magic. A task that used to take an afternoon is done in minutes. A draft appears before the coffee cools. A report that usually takes a team effort suddenly has a clear starting point.

That sense of wonder is useful. It shows you that something real has changed. But AI is not magic. There is a method underneath it, and the good news is that you do not need to build that method yourself to benefit from it.

Most small and medium businesses will never train their own model or build AI of their own. They will use it through software that already has it built in [6]. Even so, understanding the basic logic helps you evaluate those tools with confidence, recognize their limits, and tell a genuine solution from an overhyped one. The logic comes down to a few plain steps.

It starts with a clear problem. AI creates value only when it is pointed at something real, whether that is lowering customer churn, forecasting demand, or taking the manual work out of a weekly report. A vague goal produces vague results, so a little clarity at the start saves a great deal of wasted effort later.

From there it depends on data, which is just a more formal word for examples. Those examples might live in your sales records, your support conversations, or your scheduling logs. Good information leads to good output, and the reverse is just as true. Feed a tool incomplete or skewed data and its answers will be incomplete or skewed in the same way. You do not need an enormous amount of it, but you do need information that honestly reflects how your business actually runs.

With a clear problem and good examples in place, the system does its work. It looks for patterns in what it has seen, compares each new situation against them, and adjusts as it goes to sharpen its accuracy. This is the same underlying logic whether a company builds a custom model or buys a tool with one already inside. The difference is only who does the shaping. With an off-the-shelf tool, the provider has already handled the heavy lifting, and your part is to supply clean, relevant information. If you build something custom, your team or partners shape that learning with your own data. Either way, knowing how it works lets you choose the right solution and set the right expectations.

What comes back is a prediction or a recommendation, not a verdict. The system might flag a suspicious charge, surface a customer worth a phone call, or suggest the next step in a workflow. These are informed estimates, not guarantees, offered to support a decision rather than make it for you.

Which points to the part that is actually yours. Not building models or writing code, but defining the purpose, making sure your data reflects reality, choosing tools that fit, and applying the results with judgment. Get those right, and AI becomes a multiplier. Your team moves faster, misses less, follows up sooner, and has more time for the work that actually grows the business.

Leading With Understanding

That shift in mindset matters more than it first appears. The technical work can be bought. The thinking cannot.

Every time a new technology reshapes the way we work, from the assembly line to the internet, it rewards the people who lead with understanding rather than the ones who simply move fastest. The temptation today is to chase every new AI tool before deciding what it is actually for, to collect capabilities and hope value follows. But progress does not come from owning more technology. It comes from pointing the right technology at the right problem.

That begins with framing. What outcome are you really after? How does it fit the bigger picture of where the business is going? Once those questions are answered, AI stops feeling mysterious and starts becoming useful, a practical way to extend what you and your team can do.

None of this means AI is flawless. It makes mistakes, and some of them are predictable enough that a good leader should know to watch for them. We will come back to those limits later, in the chapter on what to be aware of, so you can use these tools with clear eyes rather than blind faith. For now, the takeaway is simpler. AI is not a sorcerer conjuring answers from nowhere. It is a tool, and like any tool, its value depends on the hand that guides it and the clarity behind the choice to use it.

In the next chapter, we will look at how AI and automation work together, and how to tell which one a given part of your business actually needs.

Chapter 1.2

How AI and Automation Work Together

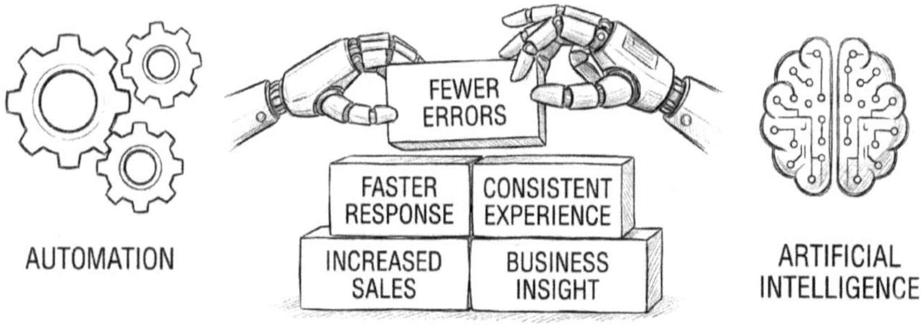

This book is, at its heart, a journey toward a single question. Where does AI genuinely make sense in your business? It is a harder question than it sounds, and you cannot answer it by looking at AI alone. To know where a new tool belongs, you have to understand the older story it is part of, the long human habit of finding better ways to work and handing off the repetitive parts. AI is the newest chapter in that story, not a break from it.

A Very Old Habit

People have always looked for ways to take the dull, repetitive work off their own hands. The plow did it. So did the printing press, the assembly line, and the humble spreadsheet. Each one took something slow and manual and made it faster, freeing people to spend their effort on what actually required them. Finding better ways to work is one of the most human things we do, and it has driven nearly all progress.

Automation is technology carrying out predefined tasks and processes with minimal human intervention. Traditional automation follows explicit rules and workflows, ensuring that repetitive and standardized activities are carried out consistently and efficiently. Examples include automated data entry, workflow routing, and system integrations.

While automation focuses on executing tasks according to predefined instructions, AI adds the ability to learn, adapt, and make decisions based on data. The combination of both technologies enables organizations to automate not only routine tasks but also more complex processes that require analysis, interpretation, and decision-making. This combination is sometimes called intelligent automation, pairing the reliability of automation with the judgment of AI.

To see how the two fit together, it helps to start somewhere familiar.

Two Different Kinds of Help

Picture making dinner the old-fashioned way. You chop every vegetable by hand, watch the oven so nothing burns, stir the sauce so it doesn't catch, and scrub the pans afterward. None of it is difficult. It is just slow, repetitive, and entirely on you.

Now picture the same evening in a modern kitchen. A food processor handles the chopping. The rice cooker shuts itself off. The dishwasher takes the cleanup. None of these appliances think about your meal. They each do one repetitive job, the same way every time, reliably. That is automation: machines carrying out fixed, predictable tasks so you do not have to.

But there is a different kind of help you might want. Imagine an assistant who knows tonight's guests, two vegetarians, one allergic to nuts, and suggests a menu that fits what is already in the fridge. That is not a fixed task done the same way every time. It is judgment, the kind no rulebook could fully cover. That is what AI adds.

You still decide what kind of evening you want. The appliances handle repetitive work. The assistant handles the thinking. Together, they make dinner easier, smarter, and better suited to the night you actually have.

The distinction the kitchen makes plain is worth stating directly. Automation follows instructions written in advance. If this happens, do that. The system is not thinking, it executes rules, step by step, exactly as written. That is its strength. In stable, predictable situations, it does the same thing every time. AI handles what you could never fully spell out in advance, the exceptions and gray areas, by recognizing patterns and making a judgment about what fits.

One line holds it together. Automation manages the structure. AI manages the judgment.

Knowing Which One a Job Needs

That is the idea. Now look at where you actually work, because the same lens applies to almost any task in your business.

Once you can see the difference between structure and judgment, you can take any process you run and split it into the part that is just rules and the part that needs thought. That split tells you where automation is enough, where AI earns its place, and where you need both. Sometimes the answer is one or the other. Often it is the two together.

Picture now a customer inquiry arriving at a busy service business, a call, a web form, a message. Part of what follows is pure structure: log it, capture the contact details, record the time, send an acknowledgment. None of that requires a moment's thought, which makes it a natural fit for automation.

But woven through that routine is a set of judgments. Is this person ready to buy, or just looking? Is it an emergency that needs a callback within the hour, or a question that can wait? What are they actually asking for, underneath how they phrased it? Who should handle it? Those depend on reading the situation, and that is where AI fits.

Sort a process this way and the path becomes clear. The repetitive structure gets automated, the judgment-heavy moments get AI, and the strongest systems are the ones where the two move as one.

What Rules Could Never Do

For decades, the software that ran a business day to day worked from rules. It could do what someone had written in advance, and it needed input in the shape it expected. Work outside the rules usually came back to a person [7].

Few examples sting like expense reports. Picture the shoebox of receipts at the end of a quarter: crumpled ones, faded thermal paper, a different layout from every store, half of them barely legible. For years, this stayed stubbornly manual. Older capture systems worked from layouts they had been set up for, so every new format meant more setup, and a shoebox from a hundred different stores was never worth it [8]. Someone keyed in the amounts and picked the categories by hand. Tedious, easy to get wrong, and almost universally despised.

That is the real limit of rules. They only handle what you anticipated, and they need their input in a shape they already recognize. The moment something varies, a new format, an odd case, a vendor you have never used, the system gives up and hands the work back to a person.

What changed is that the new layer is no longer locked into rigid templates. AI reads a receipt without needing a template. It finds the merchant, date, and amount wherever they sit, including on layouts it has never seen, though unfamiliar ones produce more errors [9]. Deciding the category is harder and less reliable. The variety that once made this expensive to automate is what it handles well, and the doubtful ones go to a person. Work that defeated rule-based systems for decades, because real life would not hold still long enough to fit a template, is now within reach.

Where the Real Power Is

Combine reliable automation with AI's judgment and something new becomes possible for a small business: running whole processes, not just isolated tasks.

Two shifts brought this within reach. The first is that the systems you already use can now talk to each other, passing information back and forth automatically through APIs, the connections that let one piece of software hand data to another, so your scheduling tool, your customer records, your inbox, and your accounting software no longer sit in separate silos that a person has to bridge by hand. The second shift matters even more. Wiring the simple ones together no longer needs a developer [10]. A new generation of tools lets you orchestrate it all visually, dragging blocks into place the way you would sketch a process on a whiteboard, instead of writing code. AI slots in wherever a step needs interpretation rather than a fixed rule.

The visual workflow manages the structure, moving information from one system to the next in order. The AI inside it manages the judgment. Reading a messy message and pulling out what matters, summarizing a long document, deciding which path to take when the answer is not black and white, even reshaping information so two systems that were never meant to talk to each other suddenly can. Work that once required an engineering team is now within reach of someone who knows their own business well. Anything more tangled still needs someone technical.

That reach is exactly why the next point matters. The same power that lets you do good work quickly can also do flawed work just as quickly and at scale. A mistake a person makes once affects a single case. The same mistake built into an automated workflow can repeat hundreds of times before anyone notices. So the discipline this book returns to again and again is this. Be clear about what you are automating and what data you are feeding it.

One principle is worth keeping in mind. Using AI does not move responsibility for a business decision onto the software provider. Your business owns how the tool is used, and the provider owns the system and the claims it makes about it. A tool can support a decision, and it can never be accountable for one.

Therefore, organizations do not necessarily need to become experts in computer science, advanced mathematics, or software development. What is essential is a clear understanding of the capabilities, limitations, and risks associated with the AI systems being used.

Controls do not have to be elaborate. The useful question is which decisions you want a person to see before they reach a customer, and what should happen when one comes back wrong. From there you check the system against numbers you already trust, keep a record of what it did, and look at it often enough to notice it drifting. None of this removes the risk. It buys you the chance to find out while the problem is still one customer rather than three hundred.

From Assistant to Something More

Notice that in everything so far, the AI has been an assistant. It exercises judgment inside the steps you laid out, but you designed the flow and you decide what happens. That is the right place to start, and it is already enough to change how a small business runs.

But there is a further step, one you can probably feel coming. What if the system did not just assist the workflow you designed, but planned and ran the whole process itself, deciding the steps and checking in only when it mattered? That is a different kind of capability, where AI stops assisting the work and begins driving it. We will come to it later in the book, once the groundwork is in place.

For now, you have the core idea: automation for the structure, AI for the judgment, and real power in the combination, claimed deliberately. The next question is the practical one. What, specifically, can AI do inside your business? In the next chapter, we will look at the handful of core capabilities that show up again and again, in nearly every organization.

The specific platforms for this work change every few months, so anything we printed here would be stale by the time you read it. We keep a live, curated list in one place instead, with templates and examples you can grab as you go.

Free Companion Toolkit

Get the Part 1 Worksheets

Scan or visit **designated.ai/books/practical-strategy**

Chapter 1.3
The Five Things AI Can Do in Any Business

Digital transformation has been around for years, yet it still feels vague to most leaders. Many confuse transformation with migration. They jump on each new technology without ever capturing the value, buying the tool while the business runs exactly as before. It usually comes down to one question too few people ask. What, specifically, do we need this to do for us? Understanding the problem has to come first.

Knowing where and how to start applying AI in your organization follows the same paradigm, and to give you some ideas, we tell the story of Innovate Corp.

The Parable of Innovate Corp

Innovate Corp was built on handshakes and hard work. For fifty years the family-owned business had made industrial valves, the kind that sit inside other companies' machines and have to work every time, and it had thrived by making them exceptionally well and treating its customers like neighbors. Arthur, the founder's son, ran the company with the same steady hand as his father, valuing loyalty and tradition above all else.

Then the world changed. Artificial intelligence was suddenly everywhere, and to the team at Innovate Corp it felt like a storm on the horizon. Smaller, tech-savvy competitors were moving at a speed that felt impossible, and the comfortable old traditions were starting to feel like anchors. After a grim quarterly meeting where they lost another long-term client to a faster rival, Arthur knew something had to change.

So he did something that raised eyebrows across the company. He hired Maya, a bright young engineer fresh out of university, and gave her a daunting but simple mandate: help them figure out this AI thing.

One: Putting Your Knowledge to Work

Innovate Corp lost the Henley account, a client of twenty years, over an avoidable mishap. Henley needed to know whether a valve could handle higher pressure in a particular setup. But Frank, the veteran who knew the answer cold, was three weeks into a long vacation. Without him, the question bounced around for over a week while a newer team member dug through scattered spec sheets and old emails. The answer that finally came back was wrong, and Henley was gone.

The wound was bigger than one client. Fifty years of knowledge about how Innovate Corp's products behaved lived nowhere the company could reach, in Frank's head, in a drawer of spec sheets, in old emails. Their greatest strength walked out the door every time someone took a vacation.

"What if anyone could just ask a question and get the answer Frank would give?" Maya said.

Sarah frowned. "Those AI chat tools only know what they learned off the internet. How would it know anything about our valves?"

The right question. A general AI chat tool is like a brilliant new hire on day one, well-read, smart, and completely ignorant of your company. Ask about your products and it will guess [11].

"That's the part we fix," Maya said. "We give it everything we know first."

The idea was straightforward. Take the documents Innovate Corp already had, the spec sheets, installation guides, service records, the technical emails

Frank had written over twenty years, and organize them so the AI could search through them when someone asked a question. Instead of guessing from general knowledge, the system would pull the relevant information from Innovate Corp's own files and use it to build an answer [12].

Within a few weeks, the team had loaded everything they could find. Sarah tested it first. She typed in the exact question Henley had asked about the valve.

The answer came back in seconds. It cited the correct pressure rating, referenced the right product series, and flagged a compatibility note from an installation report filed three years earlier. It was the answer Frank would have given, built from the same documents Frank would have checked.

"That would have saved the account," Sarah said.

Maya nodded. "That's the point. The knowledge was always here. We just couldn't get to it fast enough."

Two: The Answer Was in the Tickets

For two quarters, Innovate Corp's Net Promoter Score had been dropping, the simple measure of how likely customers are to recommend you, and no one could explain it [13]. Sales were steady. The valves were as good as ever. Over the same period, though, support tickets had climbed sharply.

Maya made the connection. More tickets meant more customers hitting a problem and writing about it. If the score was dropping, the reason was probably in those tickets.

"So we'd need someone to go through all of them and tag them," David said. "We don't have a data scientist for that, and I'm not hiring one to read complaint emails."

"We don't need one," Maya said.

What she used was a large language model, the same kind of technology behind the AI chat tools everyone had started using, but pointed at a different job. Instead of answering questions, it could read every ticket, group them by what customers were actually struggling with, and lay out the themes as a simple chart of what came up most.

First, though, the tickets needed cleaning. Names, email addresses, account numbers, anything that could identify a customer, all of it came out, replaced with simple labels like Customer 1 and Customer 2. The complaints themselves stayed exactly as written, because that was the part that mattered. What people struggled with was the signal. Who they were was not. Maya also took a minute to confirm the tool kept that data private and did not reuse it elsewhere, the kind of basic check any owner should make before handing over customer information.

Then she ran it. Work that used to require a specialist now took an afternoon, and anyone on the team could read the result.

The chart was blunt. A large share of tickets traced to two things. One installation step in the manual that kept tripping people up, and a pressure rating on the spec sheet that was easy to misread, so customers ordered the wrong part and blamed the valve. Neither problem was the product. Both were the words around the product.

The fix turned out to be the more useful discovery. The same kind of model that found the problem could help solve it. Working from what customers actually said, in their own words, the team rewrote the confusing installation step in plain language, built a short FAQ around the questions people kept asking, and redid the insert that shipped in every box. The materials stopped guessing at what customers needed and started answering what they actually asked. Within two quarters the score climbed back.

The lesson is bigger than one score. Most companies sit on a goldmine of feedback they have never truly read, support tickets, reviews, call notes, written by the only people who know exactly where the experience breaks. The volume is why no one reads it, and the volume is exactly what AI handles well. Your customers have probably already told you what is wrong. The only question is whether you are reading them one at a time, or finally seeing the whole picture they add up to.

Three: Connecting the Dots

Every month, someone on Arthur's team spent two full days building the same report by hand. The sales data lived in the CRM, the system that holds customer and deal records. The orders lived in a separate fulfillment system. The email history lived in the marketing platform. Each tool did its job well, but none of them talked to the others. So once a month, a person exported a spreadsheet from each, lined them up by hand, fixed the mismatches, and stitched them into one picture. By the time it was done, the numbers were weeks old.

The real cost was not the two days. It was every question that never got asked because answering it was too much work. Do the clients we email most actually order more? Which orders lead to repeat business? Nobody knew, because the answer lived in three systems that had no way to compare notes.

“I have more data than I’ve ever had,” Arthur said in a meeting, “and somehow I feel like I can see less.”

Maya’s fix had three parts, and none of them needed a programmer.

First, the connections. Most modern business tool comes with an API, short for Application Programming Interface. An API is a doorway the software opens so other tools can reach in and pull what they need, without anyone retyping anything. You never see it. It is worth one question before you buy any software: does it have an open API, or in plain terms, will it share its data with your other tools?

Second, something to do the connecting. Maya did not write code. She used a connector platform, a tool that links your apps together without programming, to wire the systems through their APIs: pull the customer list from the CRM, the orders from fulfillment, the email activity from the marketing platform, all into one place.

Third, something to make sense of it. The records still did not match. The same customer was “Acme Corp” in one system and “Acme Corporation” in another, with dates and account numbers formatted every which way. This was the part that used to require a data scientist and weeks of cleanup. Instead, Maya used a large language model to read the mismatched records, recognize that the two Acmes were one company, and turn the mess into a single clean picture. Then she could ask questions in plain language, the way you would ask an analyst.

So she asked the one Arthur had wondered about for years. Did the customers they emailed most actually order more?

The answer was yes, but with a twist. It was not that the email worked. It was that one message worked, and only for one group. The quarterly maintenance reminder drove heavy reorders, but almost entirely among the operations-heavy accounts that ran their equipment hard. Everyone else ignored it. And everyone got the same email.

So Maya pushed further and asked what set the reorderers apart. The analysis compared them across everything it had, order size, customer type, what they bought, which messages they opened, then reported which factors actually predicted a reorder and which were just noise. Most of what the team had assumed mattered turned out not to. One thing did. The match between the message and the kind of customer receiving it.

They had been running one playbook for a customer base that was really three or four audiences wearing the same label. The heavy-use accounts wanted uptime. The occasional buyers wanted a reason to come back at all. The same email could not serve both, so it half-served everyone.

Hence, they stopped sending one message to everyone. They split the list along the lines the data had drawn and matched each group to what it responded to. The reorder rate climbed the next quarter, and the two days of manual reporting disappeared with it.

Four: Seeing It Coming

The first thing Maya noticed on the factory floor was the waiting. A crew of four stood near a stopped line, coffee in hand, while one technician worked on a seized motor. She asked the line lead how often this happened. "Couple times a month," he said, like it was just part of the job.

Innovate Corp serviced its machines on a fixed calendar, every part swapped on schedule whether it needed it or not. Safer than waiting for a failure, but blunt, and it still did not stop the breakdowns that hit between visits.

What no one noticed was that the machines had been warning them the whole time. Modern equipment records everything as it runs: temperature, vibration, motor current, run hours. It was already being collected and written to logs nobody opened. The problem was never missing data. It was that no one could read it. Nobody watches a vibration reading drift up a fraction of a percent a week across forty machines, so the warnings piled up unread until something seized.

Reading that pattern is what machine learning does, the same engine that spots a fraudulent charge by studying millions of ordinary ones. Point it at maintenance logs and it learns what a machine's readings look like in the weeks before a failure, then watches for that shape forming again. The category has a plain name, Predictive Maintenance [2].

Here is the part that matters for using it well. The model does not tell you a motor will fail. It tells you this motor now looks like ones that failed before, so the odds are high enough to check. That is a probability, a reason to look rather than a guarantee. Some flagged machines would have run fine. A few failures will still slip through. What you get in return is a head start, trading a small number of unnecessary inspections for far fewer surprise breakdowns. That trade is worth it whenever a breakdown costs you more than an inspection, as long as you treat the flag as a reason to look and keep a person deciding what to do about it.

Using it well comes down to a few habits. Act on the warnings or the system is just noise. Track how often a flag turns out to be real, and feed that back so the predictions sharpen over time. Set the sensitivity to match the stakes. When a failure is catastrophic you want every possible warning even at the cost of false alarms, and when it is minor you raise the bar so the team is not chasing ghosts. The goal is a predictor that is right often enough to change what you do on Monday.

This reaches well beyond machinery. The signs of a customer about to leave, a project about to slip, a cash crunch six weeks out are usually already sitting in the data you keep.

Five: Winning More of the Work You Chase

Innovate Corp had three new sales reps and one veteran, Daniel, who closed at nearly twice their rate. Arthur knew Daniel was better. He could not explain why.

The usual approach was ride-alongs. Daniel would take a newer rep on a few calls, and afterward they would talk about what happened. The feedback was honest but imprecise. "You lost him when you started talking about specs." "You should have asked about their timeline earlier." Daniel was describing what he felt, filtered through instinct built over fifteen years. The newer reps nodded, took notes, and went back to making the same mistakes in slightly different ways.

The real problem was that nobody could see the call the way a tape review sees a game. By the time a rep finished a conversation, the details were already fading. They remembered how it ended, not the thirty seconds where it turned. And Daniel could not teach what he could not see in himself. His best moves were unconscious.

Maya had the team start running their calls through an AI tool that recorded, transcribed, and analyzed every conversation. What came back was not a score or a grade. It was a map of the call depicting where the customer was engaged, where they went quiet, where they started giving short answers, where the rep talked past a buying signal or answered a question nobody asked.

One pattern showed up almost immediately. The newer reps were spending the first several minutes of every call explaining what Innovate Corp did. By the time they got to the customer's problem, the energy was already gone. Daniel, without realizing it, did the opposite. He asked a question in the first thirty seconds and let the customer talk. His pitch came later, shaped around what he had just heard.

That was not a secret. It was a habit so automatic that Daniel had never thought to name it. The AI made it visible, with timestamps.

Within a few weeks the team had a small library of real moments from real calls, the ones that worked and the ones that did not. A new rep could listen to the forty-five seconds where a deal opened up and compare it against the forty-five seconds where a similar deal went flat. The lesson was concrete, not a principle in a training manual but an actual conversation they could hear and study.

The reps improved faster than classroom training had produced, because the feedback was specific, immediate, and tied to their own calls. They were not learning theory. They were watching themselves lose a deal at minute six and learning exactly what to do differently at minute six next time.

The real power showed up at scale. One call tells you what happened in that call. A hundred calls across five reps over two months tells you something else entirely. The AI surfaced patterns no single manager could have assembled. Which objections were killing deals most often? Which product questions signaled a customer was ready to buy, and how long the highest-converting calls actually lasted versus what the team assumed? It turned out that deals closed fastest when the rep spent the first third of the call asking questions and the last third on next steps. The middle was where most reps lost control, filling silence with features nobody had asked about. That insight did not come from any one conversation. It came from reading all of them together.

There was a second shift the team did not expect. Before the AI, call notes were whatever the rep remembered to type into the CRM after hanging up, usually a sentence or two, sometimes nothing. The important details, the customer's timeline, the competing bid they mentioned, the name of the decision-maker's boss, the concern they raised and then moved past, all of that lived in the rep's memory for a few hours and then disappeared. When a different rep or a manager opened that CRM record a week later, it was nearly empty.

Now the AI wrote the notes. It pulled the key information straight from the transcript into commitments made, questions left open, next steps agreed to, names and roles mentioned, objections raised, and anything the customer said about budget, timing, or competition. Every call left behind a record that was more complete than even the most diligent rep would have written, because it captured what actually happened rather than what the rep remembered to write down. A manager preparing for a follow-up or a new rep picking up an account had the full picture, not a half-sentence and a guess.

The discipline is the same one that applies everywhere in this chapter. The AI identifies patterns and captures details. It does not run the meeting. A manager still decides what matters, what to coach, and how to have the conversation. The tool makes the invisible visible. The leader decides what to do with it.

Innovate Corp's journey captures something many leaders feel today. The fear that AI might be too big, too technical, or too disruptive slowly gave way to something more productive. Not hype, not blind enthusiasm, but a cautious and genuine excitement. The kind that comes when people see real problems solved, not by replacing them, but by supporting them.

The company did not transform overnight. They did not become a tech giant. They took one meaningful step after another. Automating a painful process, amplifying a small team's creativity, and connecting data that had been scattered for years. Seeing patterns no one had noticed. Moving from reacting to predicting. Each small win reduced the fog and made the next step feel a little easier.

In the next section, we will create that spark for you. Your first win with AI is not theoretical. It can start today, and it begins with a simple, guided process that brings clarity to your most important personal or business challenge.

Let's build it together.

Chapter 1.4

Your First Win with AI

There's a version of you that exists twelve months from now.

That version has already done it. Landed the bigger clients, or the bigger role. Entered the new market, or mastered the new craft. Built the thing you keep telling yourself you'll build when things calm down. You can picture that person clearly. The road between here and there is another matter.

The Ground Is Moving

We live in a time strategists describe as volatile, uncertain, complex, and ambiguous. The military coined an acronym for it decades ago, VUCA, and it has aged uncomfortably well [14]. The skills a job asks for keep shifting, and the pressure lands hardest on smaller firms and on the roles most people assume are safest [15]. Whether you run a business or a function inside one, the ground moves while you're standing on it.

If we're being honest, AI is part of what's shaking it. New capabilities arrive monthly. Competitors move faster than they should be able to. Headlines swing between revolution and ruin. It would be strange if none of that felt unsettling.

More Information, Less Clarity

There's a second force compounding the first. The flood of information grows every year. Somewhere in that flood are the exact answers you need, which is precisely the frustration. The scarce resource is the ability to sort it. Doing that by hand, on top of a full workload, is a losing race.

Turning the Problem Into the Solution

Here's the flip. The technology stirring up the fog happens to be remarkably good at seeing through it. Sorting oceans of information, mapping gaps, laying out paths, this is precisely the work AI does well. While most people experience AI as one more source of noise, you can put it to work as an instrument of clarity. The speed of change stops being a threat the moment it's moving in your direction.

The gap between where you are and where you want to be is normal. Every goal worth having sits on the far side of one. What keeps people stuck is that the gap is unmapped. Which skills? Which experience? Which market knowledge? When you can't name what's missing, every path looks equally risky, so most of us stay busy with what's in front of us instead.

This chapter is about burning off that fog, and you're going to do it yourself, today, in about twenty minutes. You'll take an honest picture of where you stand, a clear picture of where you want to be, and let AI map what sits between them. By the last page, the question marks will have names, and the names will be arranged into a plan.

Open Your Favorite AI Chat

You'll need an AI chat assistant, and any of the major ones can handle everything this chapter asks. The leaderboard changes every few months, so naming a product here would only date this page. Our resources page keeps a current list. What matters far more than which assistant you open is what you do once it's open.

A Thirty-Second Experiment

Before the real exercise, try this. Type into your assistant exactly what most people type: "How do I grow my business?" Or, if you're focused on your career: "How do I get promoted?"

Read the answer. It will be polite, sensible, and generic. Know your customers. Improve your marketing. Communicate your value. Nothing wrong, and nothing you couldn't have written yourself. This is the moment where most capable leaders conclude AI is overrated, close the tab, and go back to work.

Here's what actually happened. You treated the chat window like a search bar, because it looks like one. What you have is closer to a very capable new hire, and no new hire does brilliant work from a one-line instruction. Hand an experienced consultant that same question on day one and you'd get the same polished generalities. Give the same consultant a real briefing, your situation, your goal, your timeline, what you need and in what form, and the output transforms. Nothing about their ability changed. The brief changed.

That's the entire secret. Asking gets you answers. Briefing gets you work. The rest of this chapter teaches you to brief.

> "The hottest new programming language is English" [16]
>
> — *Andrej Karpathy, AI researcher, 2023*

For decades, getting a computer to do real work meant learning its language. Now the machine has learned yours. Yet AI is still a machine. It can't read your face or sense what you left unsaid. It works with exactly what you give it, which is why the prompt is where everything is won or lost.

The Anatomy of a Great Prompt

A prompt is simply a written brief, instructions handed to a capable colleague. The one you're about to use looks long, and every line is doing a job. Once you understand its five parts, you'll be able to build one for anything, whether that's a marketing plan, a hiring process, or a difficult negotiation.

The fastest way to learn each part is to see what breaks without it. Every failure below will feel familiar. Those disappointing chatbot answers weren't the tool's fault. They were missing pieces of the brief.

Part 1: The Role

The prompt opens by telling the AI who to be. In our case, that's a master strategist with twenty years of experience. These systems are trained on a broad mixture of writing, from forum chatter to published research [17], and the role tells them which shelf to pull from.

Leave it out, and you get advice from nowhere, in the tone of a well-meaning friend who once read a business article. Assign it, and the answer arrives with more depth and the framing of professional advice. Careful testing shows that a role changes the character of an answer more than its correctness [18], so aim with it and still judge the content yourself.

Part 2: The Context

Next comes your situation. Where you stand today, where you want to be, and how much time you're giving yourself. This is the part most people skip, and it's the biggest reason chatbot answers feel generic. Without context, the AI can only advise some imaginary average business, because that's the only client it can see.

Give it your actual situation, and the advice starts referencing your constraints, your market, your starting point. The imaginary client leaves the room, and you take the chair.

Part 3: The Task

Then the work order. Analyze the gap. Build the roadmap. Make each step actionable. Notice that the prompt assigns work, broken into steps.

Skip this and you get the conversational equivalent of a shrug, interesting observations with nothing to do next. Sequenced steps produce sequenced thinking.

Part 4: The Constraints

Then come the rules of the game. The plan has to be realistic for a busy owner. The resources have to be specific. The projects have to produce proof. Without constraints, AI happily designs a strategy requiring three employees you don't have and a budget from someone else's business.

Constraints are how your reality gets a seat at the table. The tighter they are, the more the output belongs to you.

Part 5: The Output Format

Finally, the shape of the answer. You want a summary, a list of gaps, and a roadmap broken down by quarter, so you say exactly that. Leave this out and you'll often receive a thoughtful essay, which is a polite way of saying something you'll read once and never open again.

Demand a structure and you get a document you can act on. Thinking is half the value. The other half is a form that survives contact with Monday morning.

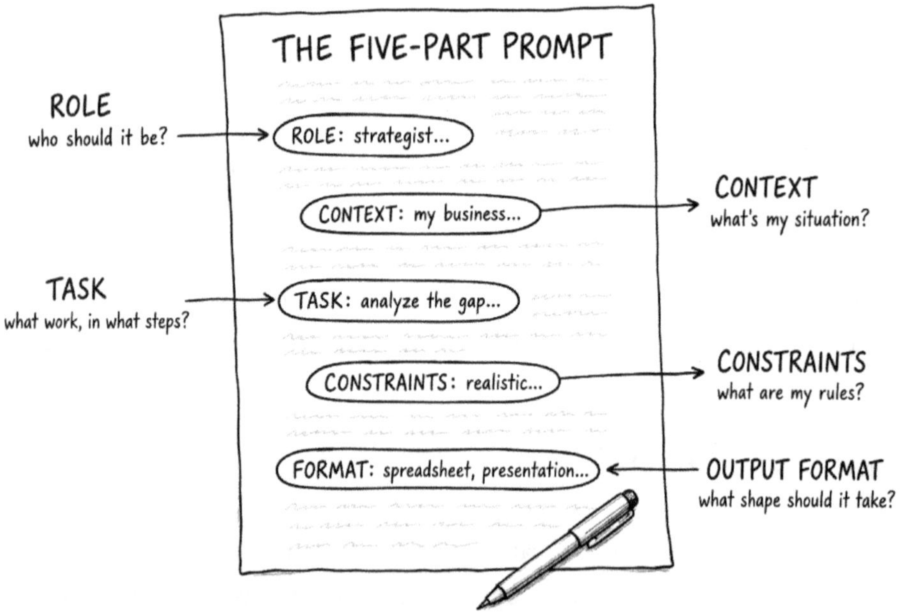

Five Parts, One Habit

Role, context, task, constraints, format. The tools will change names and get better at guessing what you meant, and the habit of briefing will still be what gets real work out of them. Nobody does their best work from a one-line instruction.

Your Turn

Time to write your first real brief. Before you open the chat, spend five minutes gathering the two ingredients that make everything else work.

First, an honest snapshot of where you stand today. If you run a business, describe what you sell, who buys it, roughly how it performs, and how the work gets done. If you're building a career, this is your resume plus the things a resume never captures, what you're genuinely good at and where you know you're thin. Write it plainly, the way you'd explain it to a trusted advisor. The AI can only see what you show it, and flattering your own profile only produces a plan for someone who doesn't exist.

Second, a specific picture of where you want to be, with a timeline attached. Vague goals produce vague roadmaps. "Grow the business" gives the AI nothing to aim at. "Reach $2 million in revenue by adding a second service line within 18 months" gives it everything. The sharper the target, the sharper the plan.

With both in hand, copy the prompt below, paste in your information where [marked], and run it.

The Personal Strategist Prompt

Role: You are a master strategist and consultant with 20 years of experience helping businesses and professionals achieve ambitious goals. You are an expert in gap analysis and in building actionable, step-by-step strategic roadmaps.

Context: I will provide three pieces of information:

1. **Current_profile:** A detailed description of my current state. This may be a description of my business, including what we sell, who our customers are, and how we operate, or my personal skills, experience, and background.

   [paste your business description or personal profile here]

2. **Target_goal:** A detailed description of the specific, measurable goal I want to achieve.

   [paste your specific goal here]

3. **Timeline:** The timeframe in which I want to achieve it.

   [enter your timeframe, for example 12 months]

TASK: Act as my personal strategist.

1. **Analyze the gap.** Compare my Current_Profile with my Target_Goal and identify the three to five most critical gaps in skills, experience, or strategy that stand between them.
2. **Build the roadmap.** Create a structured, sequential plan to close those gaps within my Timeline, broken into clear phases.
3. **Make it actionable.** For each phase, define the primary focus area, recommend specific and well-regarded learning resources, and design one practical project that produces demonstrable proof of the new capability.

Constraints: The plan must be realistic for a busy professional or business owner. Resource recommendations must be specific. Every project must generate a tangible outcome I can show, such as a portfolio piece, a new business process, or a measurable result.

Output_format: Organize your response so I can act on it immediately. Begin with a brief executive summary and the list of identified gaps. Then present the roadmap phase by phase in a clear, structured document I could transfer into a spreadsheet, a presentation, or a project plan. Close with the three actions I should take this week.

Prompt Pack

The Personal Strategist Prompt

Copy the ready-to-run prompt plus the Five-Part briefing cheat sheet.

Scan or visit **designated.ai/books/practical-strategy**

Run It, Push Back, Iterate

One more thing before you judge the result. Don't expect a final, perfect answer the first time. What comes back is a first draft from a capable strategist who just met you, and no first draft survives a real working session unchanged. The conversation is where the value compounds. If a phase looks unrealistic, say so and ask for an alternative. If a recommendation misses your market, correct it. Ask it to reorder the plan by fastest impact, or to expand the phase that worries you most. Each round sharpens the plan, because unlike a static document, this strategist revises instantly and carries no ego. Sometimes it will hold its ground instead, and those are the moments worth slowing down for.

That habit of iterating separates people who use AI from people who tried it once.

Your First Win

Take a moment to notice what just happened. Twenty minutes ago, your path forward was a set of question marks in the fog. Now the gaps have names, the names are sequenced into phases, and you know what to do this week. The kind of structured guidance that once required an expensive consulting engagement came from a well-written brief and a few rounds of honest conversation. That is your first win, and the framework behind it, role, context, task, constraints, format, now belongs to you for everything that comes next.

Next

You just watched a system produce a gap analysis and a strategic roadmap, the kind of output we associate with reasoning, judgment, and expertise. Yet at its core, the assistant you just used does something that sounds almost trivial, it predicts the next word. How does something that simple produce something that useful? The answer will change how you use them, and it's where we go next.

Chapter 1.5
How Large Language Models (LLMs) Work

Large language models have enabled widely applicable systems for text generation, summarization, question answering, information extraction, and other language-processing tasks. A machine-learning model is a parameterized mathematical representation learned from data for use in tasks such as prediction, classification, or generation. In a language model, training adjusts numerical parameters to represent statistical relationships among tokens and their contexts. If we have a closer look, a model turns out to be a mathematical representation that has learned patterns from data. For language models, these patterns capture relationships between words, concepts, and contexts, enabling the model to represent contextual relationships and generate language-like sequences. Although this behavior is often described as "understanding," that term should not imply human cognition or comprehension. Modern LLMs may contain billions of parameters distributed across multiple neural-network layers. These parameters encode statistical regularities learned during training rather than explicit human-readable facts or rules [19].

An LLM is typically one component of a broader AI system rather than a complete application by itself. Operational AI systems may integrate retrieval mechanisms, memory modules, reasoning and planning components, external tools and APIs, as well as safety and monitoring mechanisms to support reliable task execution. Before deployment, a model must be trained and evaluated. Training data must be collected, cleaned, filtered, transformed, and in some settings augmented. Evaluation then measures behavior against task-specific metrics, benchmarks, safety requirements, and acceptance criteria. Evaluation provides evidence about performance under specified test conditions, it does not by itself guarantee reliability in every deployment environment. Because foundation-model development involves large datasets and distributed computation, much of the training and evaluation pipeline is automated.

Human decisions nevertheless remain important in data governance, test design, risk assessment, and release approval. Training and serving large foundation models can require substantial computing infrastructure, energy, specialist expertise, and financial investment. For most organizations, adopting an existing foundation model and adapting it through prompting, retrieval, or parameter-efficient fine-tuning is more practical than pretraining a competitive general-purpose LLM from scratch [20].

Let us now look at an example based on my own experience.

As a systems developer and researcher, I regularly have to work with large volumes of technical documentation in order to find the information needed to solve specific problems. During the early days of prompt-based LLM applications, very few ready-to-use software solutions were available. As a result, I had to explore concepts that were mainly described in scientific literature and only used by a relatively small group of practitioners at the time.

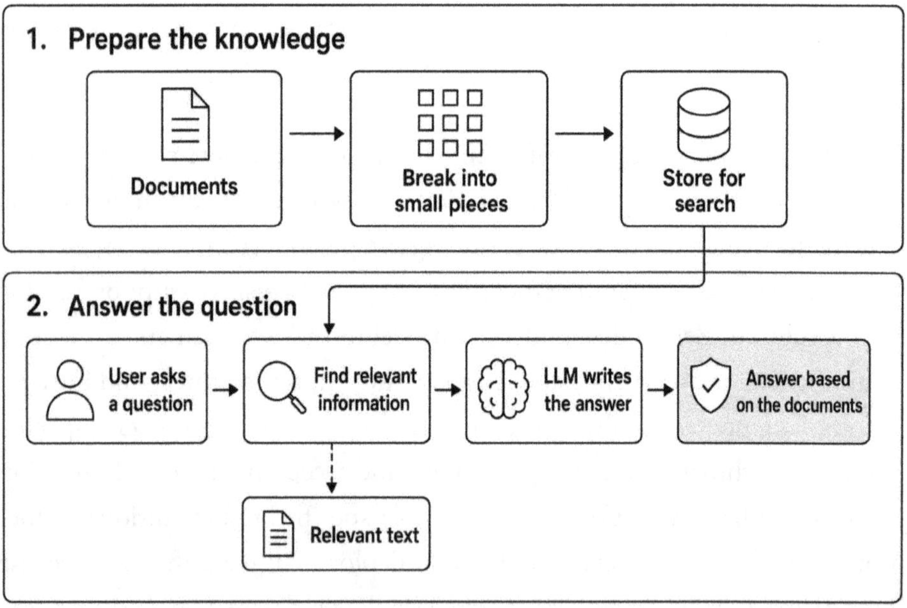

Figure 1.3 Retrieval-Augmented Generation (RAG) System

One of these concepts was Retrieval-Augmented Generation (RAG) as depicted in Figure 1.3, which combines a generative language model with an information-retrieval component. Given a request, the system retrieves potentially relevant passages from an external collection and supplies them as context to the model before it generates an answer. RAG can improve factual grounding, domain relevance, and access to information beyond a model's pretrained knowledge by retrieving relevant external information during generation. Its effectiveness depends on the quality and relevance of the retrieved information and on how that information is integrated into the generation process. Nevertheless, retrieval quality and the model's ability to effectively utilize retrieved context remain important challenges, and increasing the number of retrieved documents does not necessarily improve answer quality [21], [22].

RAG functionality is increasingly offered through commercial AI platforms that allow users to upload documents or connect repositories without implementing every retrieval component manually.

This example illustrates how retrieval-augmented systems extend the capabilities of large language models beyond relying solely on pretrained knowledge. By integrating language models with external knowledge sources, RAG enables access to up-to-date and domain-specific information during generation. Such systems support applications including open-domain question answering, conversational agents, recommendation systems, and other knowledge-intensive tasks. LLMs and retrieval-augmented systems can provide information and analysis to support human decision-making in knowledge-intensive tasks. However, their outputs remain dependent on the quality of retrieved information and the model's ability to use that information effectively [21], [22].

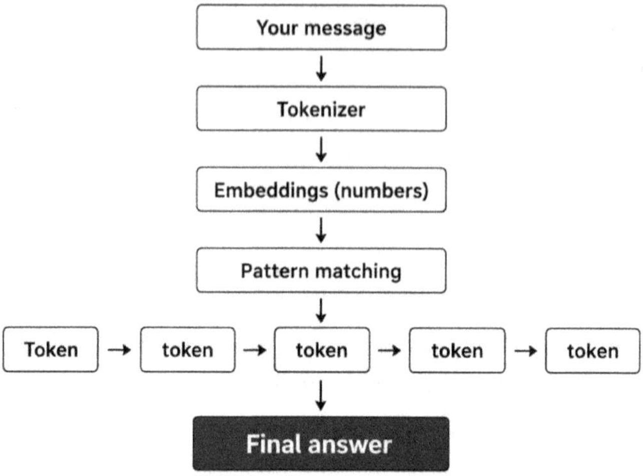

Figure 1.4 The Illusion of Reasoning. Simplified LLM generation. An LLM generates tokens using learned statistical relationships. Although its outputs can exhibit reasoning-like behavior, the extent and reliability of LLM reasoning remain active research questions.

At inference time, an autoregressive language model repeatedly estimates a probability distribution over possible next tokens and selects tokens according to a decoding strategy. Complex behavior can emerge from this process, but reasoning performance in large language models can vary substantially depending on the prompting strategy and the evaluation benchmark. Studies have shown that zero-shot, few-shot, chain-of-thought, and program-of-thought prompting can produce markedly different reasoning performance, highlighting the importance of careful evaluation design [23].

Figure 1.4 depicts an abstraction of how a large language model (LLM) works. Starting from the top, the model receives an input, which corresponds to the user's question. The input is processed by the model, and the resulting response is returned as the output.

To illustrate this concept in a practical setting, consider a manufacturing environment. Engineers and machine operators frequently need quick guidance on operating equipment, interpreting technical documentation, or troubleshooting production problems. While a general-purpose LLM can generate responses based on the knowledge acquired during training, it may not have access to company-specific documentation, the latest operating procedures, or equipment manuals. In such situations, additional mechanisms are needed to provide the model with relevant and up-to-date information.

Retrieval-augmented systems address this limitation by retrieving relevant information from external knowledge sources and providing it as contextual information for answering user queries. These capabilities make them well suited to tasks requiring access to external and domain-specific information. Research has demonstrated their application in areas including open-domain question answering, conversational agents, recommendation systems, and domain-specific knowledge retrieval. Their effectiveness depends on the quality and relevance of the retrieved information and on the model's ability to use that information effectively [21], [22].

Another potential application arises during system development. Engineers often need to search through extensive technical documentation and datasheets to locate relevant specifications and design information. An LLM-based assistant integrated with a document retrieval system can help users identify relevant information more efficiently and present it in a contextualized form. Such applications illustrate how retrieval-augmented systems can support knowledge-intensive engineering tasks involving large collections of technical documentation.

Deeper Dive

At its core, a Large Language Model has as mentioned above both an input and an output, much like any other machine learning model. The difference lies in the complexity within. LLMs are built from deep neural networks with many layers, including specialized mechanisms such as attention layers, which enable them to capture context and relationships between words. There are different architectural families of models.

Most modern large language models are based on the Transformer architecture, which employs attention mechanisms to capture contextual relationships between tokens. This architecture forms the foundation for a wide range of contemporary natural language processing applications, including text generation, summarization, question answering, translation, and information extraction [19].

Each architecture comes with its advantages and trade-offs.

For example, if a user asks an LLM about "Apple," the intended meaning depends on the surrounding context. The Transformer architecture uses attention mechanisms to capture contextual relationships between tokens, enabling the model to incorporate contextual information when generating a response. As a result, the same word can lead to different responses depending on its surrounding context [19].

Hence, it's not being smart, it's just doing math, picking words based on patterns it has seen. In other words, if the context is more technical, the output tends to be more technical as well.

It feels like magic

The emergence of large language models (LLMs) has fundamentally changed how many text-intensive tasks are performed. LLM-based systems can assist with activities such as information retrieval, summarization, classification, and drafting by processing large volumes of unstructured text and presenting relevant information in a form that is easier for users to interpret. The extent to which these capabilities translate into measurable productivity gains depends on factors such as task design, output quality, human review, system integration, and user adoption.

In organizational settings, LLMs can serve as interfaces to complex technical knowledge by transforming retrieved documents into summaries, answers, or procedural guidance. Rather than replacing human expertise, they are most effective as decision-support tools that help users access and interpret information more efficiently. Because generated outputs may be incomplete or incorrect, they should be reviewed before being used in consequential decisions.

The strategic value of AI therefore depends not only on the capabilities of the underlying models but also on their alignment with organizational needs, the availability of reliable data, appropriate governance, and clearly defined operational objectives. Organizations should evaluate AI use cases based on the value they create rather than adopting new technologies solely because they are available.

The industrial chatbot discussed in this chapter illustrates this role. By combining a language model with retrieval capabilities, such systems can provide operators and engineers with timely access to technical knowledge and guidance derived from organizational documentation. This represents an important step beyond conventional search systems. However, recent developments are extending AI even further. Instead of merely responding to user requests, emerging Agentic AI systems can plan actions, interact with external tools, and pursue goals with increasing levels of autonomy. The following section introduces this new class of AI systems and explains how it differs from traditional LLM-based assistants.

Chapter 1.6

Agentic AI from Prompts to Smart Flows

Now that we have established what an LLM is and how it essentially works, we can move on to another highly relevant and rapidly emerging technology, Agentic AI. At first, using AI felt revolutionary e.g. type a question, get an instant answer. Need an email? A summary? A plan? Just ask, and the magic happens. But after the excitement faded, something became clear, these systems only responded.

Agentic AI describes systems that pursue objectives through iterative observation, planning, tool use, action, and evaluation. In these architectures, an LLM typically provides language processing and reasoning capabilities, while complementary modules manage memory, interact with external tools, maintain execution context, and incorporate feedback.

Unlike a standalone prompt–response model, an agentic system integrates these capabilities to perform actions in external systems in pursuit of defined goals [24].

It connects information, automates steps, and moves work forward without waiting for every instruction. This is Agentic AI, the evolution from prompt-based assistance to smart, self-directed workflows that turn ideas into results.

Agentic AI refers to artificial intelligence systems that can autonomously take action toward goals, not just respond to user prompts. Unlike traditional AI chatbots that wait for instructions, Agentic AI can understand context, make decisions, and execute multi-step tasks on behalf of the user, often coordinating with tools, data, and other systems.

The heart of Agentic AI lies in autonomy, context-awareness, and action.

Agentic AI systems may decompose objectives into intermediate tasks, select tools, invoke APIs, retrieve information, and revise plans using observed results [24]. Furthermore, agentic systems can maintain and use representations of task context, goals, previous interactions, and tool outputs. This does not necessarily imply human-like understanding.

It represents the shift from "AI as a conversational assistant" to "AI as a proactive collaborator."

Rather than stopping at generating text or answering questions, an Agentic AI can plan what needs to be done to achieve a goal, or act by using APIs, software, or workflows to complete steps automatically. Some agents adapt their next action using feedback from earlier steps. Persistent learning across separate executions occurs only when the system includes an explicit learning or memory-update mechanism.

In other words, Agentic AI doesn't just help you think, it helps you get things done. It turns prompts into processes, and conversations into continuous workflows. When people hear the word "agent", they often think of a travel agent, a real estate agent, or maybe a call center agent. So, a person who represents you and acts on your behalf. In the world of technology, the word "agent" has taken on a new meaning. An AI agent is not a person at all, but a piece of software designed to observe, decide, and act in pursuit of a goal. Think of a chatbot answering questions, or the navigation system that reroutes your car when there's traffic ahead.

So, when we talk about "Agents versus Agents", here's what we mean. Human users and AI agents provide complementary capabilities. Agentic AI systems can decompose objectives into intermediate tasks, select and invoke external tools, retrieve relevant information, and iteratively revise their plans based on observed outcomes. Rather than operating as standalone prompt–response models, these systems combine language models with memory, planning, and tool-integration components to support the completion of complex, goal-oriented tasks [24].

Imagine an airplane where the autopilot system is like an AI agent. It reads the data, keeps the plane steady, and works without fatigue. On the other hand, the pilot is like a human expert. They bring judgment, context, and the ability to handle unexpected situations, like a sudden storm or an emergency. Please, be aware that current LLMs do not have aviation-grade assurance.

Both are important. The autopilot provides efficiency and consistency, while the pilot ensures safety, adaptability, and trust. Hence, AI agents are not here to replace human experts, but to work alongside them. By handling the routine and repetitive tasks, AI frees people to focus on the complex, creative, and relationship-driven work for which human judgment, accountability, empathy, or contextual understanding remains essential.

Summary:

- AI agents = speed, scale, and automation.
- Human experts = empathy, judgment, and adaptability.

The future is about how the two can complement each other to achieve more together instead of one versus the other. Most people today see AI as a tool that helps with single tasks. You ask it to "write this email", "summarize this article", or "fix this sentence." That's useful, but it's only scratching the surface. An AI agent goes a step further. Rather than waiting for you to hand it to each little job, it can take on a bigger responsibility. A good example is if you tell the AI agent to "handle my inbox", instead of just writing one email.

An agentic AI system could assist with email management by classifying incoming messages, retrieving relevant information, drafting routine responses, and suggesting appropriate follow-up actions. This illustrates how agentic systems combine language models with planning, memory, and external tool integration to support the completion of goal-oriented tasks [24]. In other words, agents move from being assistants for tasks to partners for outcomes. This shift is what makes AI agents so exciting. They're not just doing what you ask, they're taking on goals, managing workflows, and freeing you up to focus on what really matters.

To start with agents you need to understand the concept first. An agent is essentially a computer program which works on a problem you define initially. The agent gets a role which you give it as well. That positions the agent into an area being an expert solving your problem. Then you need to tell your agent what the problem actually is. Further, you define which data sources are allowed and appreciated to work with and how the agent should behave. The last point is important. Don't let the agent decide how to work through your computer and how to use your local infrastructure. Finally, start your agent. That's essentially it. A common abstraction represents an agent as a feedback loop. It receives observations, updates its internal state, selects or plans its next step, executes it through an available tool, and observes the result. Real systems may include separate modules for planning, memory, retrieval, tool selection, execution, and feedback, which together enable the completion of complex, goal-oriented tasks [24].

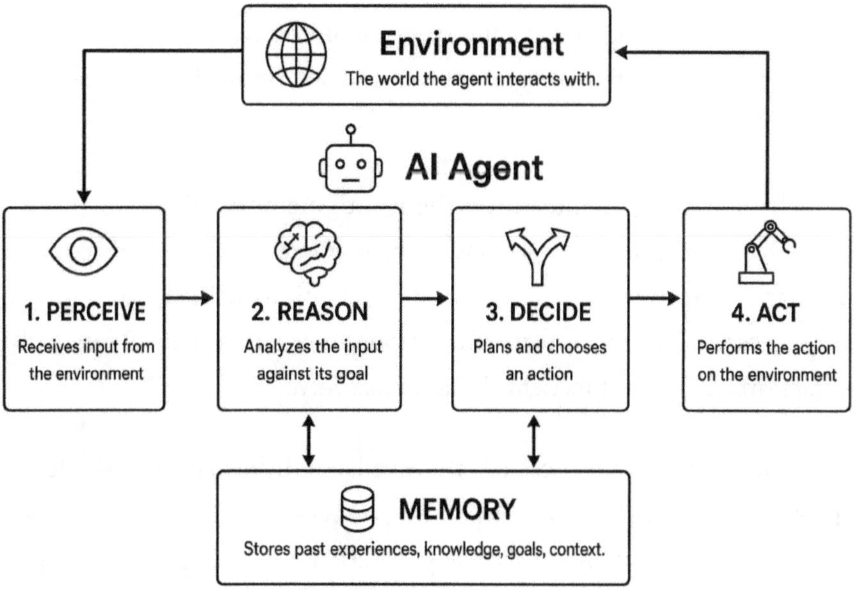

Figure 1.5 How an AI-Agent looks like in principle. The agent continuously loops: it percieves the environment, processes,plans and selects an action using memory, acts, and then percieves the updated environment again.

If you look at the figure above (see Figure 1.5), you can see a simplified representation of how an AI-agent operates. Although real-world AI agents can be considerably more sophisticated, the illustration captures the fundamental components and workflow common to most agentic systems. An AI agent continuously interacts with its environment through a closed perception-action loop. First, it perceives information from its environment, such as sensor data, user requests, documents, databases, or other digital resources. This information forms the basis for the agent's internal reasoning process.

At the center of the architecture is the AI agent itself. While a Large Language Model (LLM) often provides the core language understanding and reasoning capabilities, the agent consists of much more than the model alone. It also includes components responsible for perception, reasoning, planning, decision-making, memory management, and action execution. These components work together to interpret incoming information, decompose complex objectives into manageable tasks, and determine the most appropriate actions.

A key element of the architecture is the agent's memory, which stores relevant knowledge, previous interactions, intermediate results, and contextual information. By combining its current observations with stored knowledge, the agent can make more informed decisions, maintain context across multiple interactions, and adapt its behavior as tasks evolve.

Once the agent has processed the available information, it selects an appropriate course of action. Depending on its capabilities and permissions, these actions may include accessing external knowledge bases, invoking software applications, calling APIs, controlling physical devices, or interacting with other AI services.

This ability to combine reasoning with external tool usage enables AI agents to solve problems that extend well beyond traditional question/answering systems. The interaction follows a continuous feedback loop. After executing an action, the agent observes the resulting changes in the environment, evaluates whether progress has been made toward the intended objective, and determines the next appropriate step.

If necessary, it revises its plan, gathers additional information, or performs further actions until the desired outcome has been achieved. Consequently, an AI agent can be viewed as a goal-oriented system that continuously perceives, reasons, decides, acts, and learns from the resulting feedback. This iterative decision-making cycle is one of the defining characteristics of Agentic AI and enables the automation of complex workflows that require multiple coordinated decisions and interactions over time.

Of course, the concept of an "agent" is not entirely new. Traditional software systems have been performing automated tasks for decades. For example, a simple program can continuously monitor an email inbox, classify incoming messages, and store them in the appropriate location. Such systems follow predefined rules and workflows that have been explicitly programmed by developers.

Modern AI agents, however, represent a significant evolution of this concept. Rather than relying solely on fixed rules, they leverage Large Language Models and other AI technologies to interpret instructions, reason about problems, adapt to changing situations, and determine appropriate actions. This flexibility allows them to handle tasks that would be difficult or impractical to solve using traditional programming approaches alone.

When people talk about AI agents today, they are usually referring to these intelligent, LLM-powered systems rather than conventional automation scripts. Their ability to understand natural language, access external tools, retrieve information, and execute multi-step workflows is what makes them particularly powerful.

Fortunately, creating such agents no longer requires deep expertise in software development or low-level programming. A growing number of platforms provide user-friendly interfaces that allow users to configure, test, and deploy AI agents through graphical tools rather than writing extensive amounts of code. These platforms make Agentic AI accessible to a much broader audience and enable organizations to rapidly develop solutions for a wide range of business scenarios.

Chapter 1.7

Limits to Be Aware Of

In the earlier chapters we made the case for AI as a capable partner in our daily work. It generates, summarizes, and forecasts on the spot, and there is no doubt it takes a real load off our shoulders.

These tools also keep improving. Today they are more capable than they were a year ago, and a year from now they will be better still. That matters for this chapter, because it means the limits we are about to discuss are a moving target. Some will shrink as the technology matures. Others are inherent to how AI works and will be with us for a long time.

This chapter looks at two kinds of limits. The first comes from the technology itself, from the way these models actually work. The second is more nuanced. As the tools get easier to use, the difference between using them well and leaning on them too hard starts to show in the work. We will take the technology first.

Common AI Model Issues

Every AI model has limits, and they come from two places: the data it learned from and the way the technology works. The four that follow, bias, overfitting, hallucination, and lost context, are the ones you are most likely to meet, and each one is easier to manage once you can recognize it.

Bias: Skewed In, Skewed Out

Bias in an AI system refers to systematic tendencies that can lead to unfair or undesirable outcomes, such as favoring particular groups or decisions. These biases may arise from the training data, the model development process, or other aspects of system design and deployment [25]

If the idea of bias sounds familiar, that is because it is. Every leadership program teaches it. We are all shaped by our own experience, and it affects our judgment in ways we do not always see. Good leaders learn to catch it in themselves. AI inherits the same flaw for the same reason. It learns from past examples, exactly the way we do, so when the examples it learned from were skewed, its judgment comes out skewed too. The difference is that a biased person at least sounds like an opinion. A biased model sounds like a fact.

To see how that happens, you have to know how a model learns. You train it on data, which is just a large collection of past examples. Those examples can be almost anything: thousands of resumes, years of loan decisions, customer service chats, medical images, sales records. The model studies that collection, finds the patterns inside it, and uses them to judge each new case it meets. If the examples lean one way, for any reason, the model learns to lean the same way, and it keeps leaning long after anyone remembers why. Skewed data going in produces skewed answers coming out.

AI now helps decide who gets approved for a loan, who lands the interview, who gets into the program. When a hidden bias rides along into decisions like those, it stops being a technical footnote and starts shaping real outcomes for real people.

Walk through how it happens in a setting most leaders know well, hiring. Say you want to speed up screening, so you train a model to spot strong applicants. You give it the most natural teacher you have, ten years of your own hiring decisions, every resume you received and who you ended up hiring.

Now look at what the model actually learns from that. It does not learn who is good at the job, because it cannot see job performance. All it sees is who got hired, so that is the pattern it copies. If the people hired over those ten years came mostly from a few companies, leaned one way in certain roles, or shared a similar background, the model reads that as the recipe for a good candidate.

Then you feed it with new applicants. A strong one comes through who does not fit the old mold, a different background, a non-traditional path, a gap in the timeline. The model scores them low. They simply do not look like the people who were hired before. The model is doing exactly what you trained it to do. It is repeating your history.

The clearest proof comes from controlled tests. In one study, researchers took thousands of mortgage applications and ran them through an AI system, changing only one thing: the applicant's race. The financial details were identical. The credit scores were identical. The only difference was a demographic label, and the recommendations still came back tilted, with some applicants needing far stronger numbers to earn the same approval as others.

A model trained on past data will carry the patterns in that data forward, fair or not. For legal and ethical reasons both, you need to understand this before you rely on a model's judgment, and even more so if you are building or training one on your own data. It is our responsibility to understand the risk of bias and to make every effort to keep it out of our models.

Overfitting: Too Good to Be True

When my journey with AI started, I enrolled in an applied data science course at MIT. For my capstone I chose the medical field, partly because it was completely new to me and partly because it felt like something that mattered. The project was to build a model that could detect malaria from cell images.

With each new version of the model, I tracked one number, how accurately it identified the infected cells. It started around 85 percent, climbed into the low nineties, and when it reached 98 the excitement turned to suspicion. The number was too good. The model was not getting smarter at spotting malaria. It was overfitting, learning the training images so tightly that it looked nearly perfect on the cells it had already seen and would stumble the moment it met a new one. I turned out to be right. It scored beautifully on the data it trained on and would have done poorly on real samples. This was malaria, where a confident wrong answer can cost someone their life.

That is overfitting, with the tendency of a model to fit the training data too closely, so that it performs well on the data it has seen during training but less effectively on new, unseen data [26].

It usually starts with the person building the model. When you build one, you choose which details it should pay attention to, the signals it uses to make a call. These are called features. Give it too many, or feed it examples that all look alike, and the model latches onto every little quirk in the training data, including the random coincidences that mean nothing. Those quirks fit the examples it studied and describe nothing in the real world. So it looks flawless on what it trained on and falls apart the moment something new arrives. It was memorizing the noise instead of learning the real patterns, the ones that actually hold up when new data comes in.

A navigation app shows the harmless version. It learns your morning commute down to the second, then sends you in circles the first day there is roadwork, because the route it memorized was built for a world that no longer exists.

The same flaw hides inside everyday business tools. A sales forecast tuned tightly to a few steady years can look sharp right up until the market shifts, then keep predicting the old world with the same confidence it always had. It learned what demand happened to look like in the years it studied, and it cannot tell the difference.

So how do you guard against it? If you are building or training a model, the basic discipline is to never judge it on the data it learned from. Hold some examples back, cases the model never sees while training, and test it on those. If it scores high on what it studied and low on what it did not, you have caught overfitting before it reached the real world. That gap between the two scores is the single most useful thing to watch. Resisting the urge to pile on more and more features helps too, since a simpler model is often the one that generalizes better.

Most leaders buy AI tools rather than build them, and you cannot see inside a tool you buy. Stay curious and ask for details. How was it tested, and on what data. Was it proven on fresh real-world cases or only on the examples it trained on. How does it hold up when conditions shift? These are the differences between a tool that works in your business and one that only works in a slide deck. Once it is in your hands, stay attentive and keep testing it against your own reality, so a small problem in the demo does not become a large one at scale.

Hallucination: Confidently Making Things Up

You may have come across the term "hallucination." It is commonly used to describe situations in which a large language model generates responses that are incorrect or non-factual. Although LLMs are capable of producing fluent and convincing text, they are not always reliable and may generate information that is unsupported by evidence or inconsistent with the source material. Such unfaithful responses are referred to as hallucinations and can pose significant risks, particularly in safety-critical applications. A response that sounds authoritative and is simply made up. The model may cite a study that was never written, quote a source that does not exist, or state a number it essentially guessed [27].

The mistakes are often small and specific, which is what makes them slippery. Ask for a summary of a report and the model may hand back a clean summary along with a fact about the author that appears nowhere in the document. Ask for statistics and it may produce numbers that look reasonable and trace back to nothing. Ask it to recommend sources and it may invent titles, authors, and dates that fit the shape of a real reference without being one. The output is not random nonsense. It is plausible, well formed, and wrong, which is the hardest kind of error to catch.

Lawyers have learned this in public. In a now-famous case, attorneys filed a court brief built on cases an AI tool had supplied. The citations looked real, with plausible names, judges, and quotes. They were entirely invented. The lawyers, trusting how convincing the output sounded, never verified it. The court did. The result was fines, public embarrassment, and lasting damage to their standing. What turned this from a single mistake into a pattern is the volume. Courts have now flagged well over a thousand filings worldwide for AI-invented citations, and the number keeps climbing.

Here is the part worth sitting with. The hallucination itself is harmless until a person acts on it. The model producing a fake citation costs nothing. A human copying it into a legal brief without checking is where the damage begins. That is the thin line every user walks, the pull between being efficient and being effective. Skipping the check is faster, and the tool is right often enough that skipping it usually gets away with it. But fast and wrong is not effective. The more you lean on the speed, the easier it becomes to stop verifying, and that is exactly when a confident hallucination sails straight through into something that matters.

So the guard is a habit, not a one-time fix. Treat anything a model gives you that carries a fact, a figure, a quote, a name, or a citation as a claim to be checked, not a truth to be trusted. The higher the stakes, the harder you check. Verify against a source you already know is reliable before the output leaves your hands or shapes a decision. The model is a fast and articulate assistant. It is not a witness under oath.

Context: What the Model Does Not Know

With generative AI, context is the information you have put in front of it, e.g. your messages, instructions, and any files you have shared. That is all it has to work with. Anything you have not given it, it does not know, and it will not tell you what it is missing. This shows up in two ways.

First, it does not know your business. It has never seen your margins, your customers, or the way things really work at your company. So its advice can sound reasonable and still be wrong for your situation. Ask it to book a meeting for when you land at three o'clock, and it may schedule it for three, never accounting for the time to get off the plane and reach the room. It read the words. It missed the situation.

Second, it forgets. A generative tool holds only so much of a conversation at once, and if you pass that limit, the earliest parts drop away. Picture building a piece of software with an AI assistant. For four or five hours it works beautifully, until it loses track of what it did earlier in the session and starts rewriting the very code that was already working, breaking things it had fixed hours ago. The early work has fallen out of its view, so it no longer knows that code exists.

The fix for the first problem is to give the tool the context yourself, up front and in detail. The fix for the second is to stop keeping important context only in the chat. Keep a running note somewhere you control that captures what you are building, the decisions you have made, what is working, and where you left off. The simplest place is a plain text file. Many people use a format called markdown, saved with a ".md" extension, which is just a text file with a couple of light symbols for structure, a pound sign for a heading, a dash for a list item.

It stays readable to you and easy for the tool to read back. Keep that file updated, and when a conversation forgets or you start a new one, you can bring the tool back up to speed in seconds instead of losing hours.

When Easy Becomes a Problem

Generative AI does not change who you are. It magnifies it. Whatever you bring to the work, it makes more of.

If you are in a hurry and happy with good enough, AI lets you produce sloppy work faster than ever, and that sloppiness shows. If you are careful but you lean on the tool for everything and add nothing of your own, you get the opposite failure, work that comes out polished, generic, and hollow. It looks fine at a glance and says nothing. Any sharp person reading it can tell that no human really thought it through, and that lands as an insult, a signal that their time was not worth your effort.

I learned this from both sides, and the first time, I was the one who got it wrong.

Just a few months ago, a co-founder who leads software architecture on one of my ventures asked me for some documentation. I sat down with an AI tool and worked at it for hours, shaping and expanding, and I was proud of what came out. He was not. He was annoyed that I had so obviously leaned on AI to answer him.

I got defensive. I decided the problem was him, that he was closed off and not forward-looking about technology. He started using AI to generate his replies to me, and for a while the whole thing soured.

We worked through it. But the lesson did not fully land until later, when my own inbox began filling with AI-generated emails from other people. Long, polished, and empty. Answers with no substance, no specifics, nothing that showed a person had actually engaged with my questions. Every one of them irritated me, and it was the same irritation my co-founder must have felt. It reads as a lack of respect and a waste of my time. I had done to him exactly what these emails were now doing to me.

That is the part worth taking seriously. As AI gets easier and more people lean on it for everything, this kind of work is spreading hollow documents, templated presentations, emails padded with noise. It speaks poorly of the person who sends it, whether they meant it to or not. Standards are getting harder to hold, and the people who let theirs slip will pay for it in how others come to see them.

Confidence Without Competence

The new wave of AI assistants, the copilots that sit beside us as we work, can now do things most of us never knew how to do. They write the code, build the spreadsheet, draft the contract, screen the applicants. That is genuinely powerful. It is also where a particular trap opens up, because being able to produce something is not the same as understanding it, and the tool will not tell you the difference.

I have watched this play out. A business owner, brilliant in their own field but with no background in code, decided on their own to vibe code a new intake form onto their website, letting an AI assistant write it step by step while they steered on. Each step seemed to work, and they felt like they had it. They were proud of it, right up until the site went down. Then came the part that really cost them. They could not tell what broke, or even that something had broken, because the same gap in knowledge that caused the problem hid it from them. The tool had given them a developer's output without a developer's eye, the instinct that something is wrong before a customer ever notices. Nobody caught it for days, and a site down for days bleeds thousands of dollars in lost business and wasted ad spend that never comes back.

The same trap shows up in smaller, everyday tasks. An employee at one of my client companies recently shared what they considered a real AI win. They fed a stack of resumes into a chatbot, asked it to find the best people, and passed along the three top candidates it picked. They had no experience with prompting and no sense of what the tool could and could not do. Some of those resumes were almost certainly scanned images the tool could not read, which means it judged the field without ever seeing part of it. They simply trusted a confident answer from a tool whose limits they did not understand, and unless someone knew to check, no one would ever learn that good candidates had been dropped for the wrong reason.

The through line is simple. These tools let us reach past our own expertise, and that is a gift. It is also a risk, because they hand us the results of a skill we do not have, and without that skill we cannot judge whether the results are any good. Feeling capable is not the same as being capable, and the gap between them is exactly where the trouble hides.

Human in the Loop

Look back at everything in this chapter and one thread runs through all of it. Bias, overfitting, hallucination, lost context, those are the technology's limits. Sloppy work, false confidence, leaning too hard, those are ours. Different sources, one remedy. Keep a human in the loop.

The phrase means more than having a person nearby. It means someone who oversees the work, stays curious about how the tool reached its answer, and holds the authority to override it. These systems are built by people and used by people, and both roles need that same person. The bias enters when whoever built the model never asked who the data left out. The hallucination does its damage when whoever used it never checked the citation. The hollow email goes out when no one stops to shape it. At every point, someone should be watching, and watching with a critical eye, not a trusting one.

The good news is that this is within anyone's reach. You do not need to understand the mathematics inside a model to use it well. You need to stay engaged, stay a little skeptical, and apply your own judgment to whatever the tool hands back. Use AI for what it is, a fast and capable assistant that extends what you and your team can do, then do the part that was always yours. Read it, question it, weigh it against what you know, and make the call. AI will keep getting better, faster, and more convincing, and none of that removes the need for human intelligence and plain common sense. The more capable these tools become, the more it matters that a thinking, responsible person stays in the loop, using them to do the right things well.

References

1. Momin E, Cook T, Gershon G, Barr J, De Cecco CN, Van Assen M. Systematic review on the impact of deep learning-driven worklist triage on radiology workflow and clinical outcomes. Eur Radiol. 2025 May 21;35(11):6879–93. doi:10.1007/s00330-025-11674-2

2. ISO. Information technology. Artificial intelligence. Artificial intelligence concepts and terminology [International Standard]. Geneva: ISO; 2022.

3. Ghosh, Reilly. Credit card fraud detection with a neural-network. In: Proceedings of the Twenty-Seventh Hawaii International Conference on System Sciences HICSS-94 [Internet]. Wailea, HI, USA: IEEE Comput. Soc. Press; 1994 [cited 2026 Aug 3]. p. 621–30. Available from: http://ieeexplore.ieee.org/document/323314/ doi:10.1109/HICSS.1994.323314

4. Zolas N, Kroff Z, Brynjolfsson E, McElheran K, Beede D, Buffington C, et al. Advanced Technologies Adoption and Use by U.S. Firms: Evidence from the Annual Business Survey [Internet]. Cambridge, MA: National Bureau of Economic Research; 2020 Dec [cited 2026 Aug 3]. p. w28290. Report no.: w28290. Available from: http://www.nber.org/papers/w28290.pdf doi:10.3386/w28290

5. Jurafsky D, Martin JH. Speech and Language Processing: An Introduction to Natural Language Processing, Computational Linguistics, and Speech Recognition with Language Models [Internet]. 3rd ed. 2026. Available from: https://web.stanford.edu/~jurafsky/slp3/

6. European Commission. Eurostat. The use of artificial intelligence (AI) technologies in the European Union [Internet]. LU: Publications Office of the European Union; 2026 [cited 2026 Aug 4]. (Statistical

reports). Available from: https://data.europa.eu/doi/10.2785/9221093 doi:10.2785/9221093

7. Land F. The first business computer: a case study in user-driven innovation. IEEE Ann Hist Comput. 2000 Sep;22(3):16–26. doi:10.1109/85.859523

8. Hamza H, Belaïd Y, Belaïd A. Case-Based Reasoning for Invoice Analysis and Recognition. In: Weber RO, Richter MM, editors. Case-Based Reasoning Research and Development [Internet]. Berlin, Heidelberg: Springer Berlin Heidelberg; 2007 [cited 2026 Aug 4]. p. 404–18. Available from: http://link.springer.com/10.1007/978-3-540-74141-1_28 doi:10.1007/978-3-540-74141-1_28

9. Šimsa Š, Šulc M, Uřičář M, Patel Y, Hamdi A, Kocián M, et al. DocILE Benchmark for Document Information Localization and Extraction. In: Fink GA, Jain R, Kise K, Zanibbi R, editors. Document Analysis and Recognition - ICDAR 2023 [Internet]. Cham: Springer Nature Switzerland; 2023 [cited 2026 Aug 4]. p. 147–66. Available from: https://link.springer.com/10.1007/978-3-031-41679-8_9 doi:10.1007/978-3-031-41679-8_9

10. Guthardt T, Kosiol J, Hohlfeld O. Low-code vs. the developer: An empirical study on the developer experience and efficiency of a no-code platform. In: Proceedings of the ACM/IEEE 27th International Conference on Model Driven Engineering Languages and Systems [Internet]. Linz Austria: ACM; 2024 [cited 2026 Aug 4]. p. 856–65. Available from: https://dl.acm.org/doi/10.1145/3652620.3688332 doi:10.1145/3652620.3688332

11. Ji Z, Lee N, Frieske R, Yu T, Su D, Xu Y, et al. Survey of Hallucination in Natural Language Generation. ACM Comput Surv. 2023 Dec 31;55(12):1–38. doi:10.1145/3571730

12. Fan W, Ding Y, Ning L, Wang S, Li H, Yin D, et al. A Survey on RAG Meeting LLMs: Towards Retrieval-Augmented Large Language Models. In: Proceedings of the 30th ACM SIGKDD Conference on Knowledge Discovery and Data Mining [Internet]. Barcelona Spain: ACM; 2024 [cited 2026 Aug 4]. p. 6491–501. Available from: https://dl.acm.org/doi/10.1145/3637528.3671470 doi:10.1145/3637528.3671470

13. Reichheld FF. The One Number You Need to Grow. Harvard Business Review [Internet]. 2003 Dec;81(12). Available from: https://hbr.org/2003/12/the-one-number-you-need-to-grow

14. U.S. Army Heritage and Education Center. Who First Originated the Term VUCA (Volatility, Uncertainty, Complexity and Ambiguity)? [Internet]. Available from: https://usawc.libanswers.com/ahec/faq/84869

15. Tong D, Wu L, Evans JA. Lower-skilled occupations face greater upskilling pressure in U.S. job ads. Nat Commun. 2025 Dec 31;17(1):1237. doi:10.1038/s41467-025-67992-y

16. Karpathy A. The hottest new programming language is English [Post]. X, formerly Twitter [Internet]. 2023. Available from: https://x.com/karpathy/status/1617979122625712128

17. Soldaini L, Kinney R, Bhagia A, Schwenk D, Atkinson D, Authur R, et al. Dolma: an Open Corpus of Three Trillion Tokens for Language Model Pretraining Research. In: Proceedings of the 62nd Annual Meeting of the Association for Computational Linguistics (Volume 1: Long Papers) [Internet]. Bangkok, Thailand: Association for Computational Linguistics; 2024 [cited 2026 Aug 5]. p. 15725–88. Available from: https://aclanthology.org/2024.acl-long.840 doi:10.18653/v1/2024.acl-long.840

18. Zheng M, Pei J, Logeswaran L, Lee M, Jurgens D. When "A Helpful Assistant" Is Not Really Helpful: Personas in System Prompts Do Not Improve Performances of Large Language Models. In: Findings of the Association for Computational Linguistics: EMNLP 2024 [Internet]. Miami, Florida, USA: Association for Computational Linguistics; 2024 [cited 2026 Aug 5]. p. 15126–54. Available from: https://aclanthology.org/2024.findings-emnlp.888 doi:10.18653/v1/2024.findings-emnlp.888

19. Raiaan MAK, Mukta MdSH, Fatema K, Fahad NM, Sakib S, Mim MMJ, et al. A Review on Large Language Models: Architectures, Applications, Taxonomies, Open Issues and Challenges. IEEE Access. 2024;12:26839–74. doi:10.1109/ACCESS.2024.3365742

20. Zhou J, Chen Y, Hong Z, Chen W, Yu Y, Zhang T, et al. Training and Serving System of Foundation Models: A Comprehensive Survey. IEEE Open J Comput Soc. 2024;5:107–19. doi:10.1109/OJCS.2024.3380828

21. Shahade AK, Deshmukh PV. Enhancing Natural Language Processing: A Comprehensive Review of Retrieval Augmented Generation. In: 2024 4th International Conference on Sustainable Expert Systems (ICSES) [Internet]. Kaski, Nepal: IEEE; 2024 [cited 2026 Aug 5]. p. 609–11. Available from: https://ieeexplore.ieee.org/document/10763224/ doi:10.1109/ICSES63445.2024.10763224

22. Liu NF, Lin K, Hewitt J, Paranjape A, Bevilacqua M, Petroni F, et al. Lost in the Middle: How Language Models Use Long Contexts. Trans Assoc Comput Linguist. 2024 Feb 23;12:157–73. doi:10.1162/tacl_a_00638

23. Xu F, Lin Q, Han J, Zhao T, Liu J, Cambria E. Are Large Language Models Really Good Logical Reasoners? A Comprehensive Evaluation and Beyond. IEEE Trans Knowl Data Eng. 2025 Apr;37(4):1620–34. doi:10.1109/TKDE.2025.3536008

24. Acharya DB, Kuppan K, Divya B. Agentic AI: Autonomous Intelligence for Complex Goals—A Comprehensive Survey. IEEE Access. 2025;13:18912–36. doi:10.1109/ACCESS.2025.3532853

25. Harris CG. Combining Human-in-the-Loop Systems and AI Fairness Toolkits to Reduce Age Bias in AI Job Hiring Algorithms. In: 2024 IEEE International Conference on Big Data and Smart Computing (BigComp) [Internet]. Bangkok, Thailand: IEEE; 2024 [cited 2026 Aug 5]. p. 60–6. Available from: https://ieeexplore.ieee.org/document/10488147/ doi:10.1109/BigComp60711.2024.00019

26. Li Z, Kamnitsas K, Glocker B. Analyzing Overfitting Under Class Imbalance in Neural Networks for Image Segmentation. IEEE Trans Med Imaging. 2021 Mar;40(3):1065–77. doi:10.1109/TMI.2020.3046692

27. Malin B, Kalganova T, Boulgouris N. A Review of Faithfulness Metrics for Hallucination Assessment in Large Language Models. IEEE J Sel Top Signal Process. 2025 Oct;19(7):1362–75. doi:10.1109/JSTSP.2025.3579203

Part 2

AI STRATEGY

Chapter 2.1
The AI Trap: Adoption Without Alignment

Before we go any further, I want to start with something honest. While this may sound negative, it comes from a place of optimism.

I am genuinely excited about what AI can do in our time. I have seen how it helps small teams move faster, gives leaders clarity, and unlocks creativity in people who never saw themselves as technical.

But to get there, we need to acknowledge the reality first. Not to discourage anyone, but to clear the noise and make space for what is truly possible. So bear with me. We will start with the hard truths, then shift to the exciting part: what AI can actually do for your business today.

The Trap: Activity That Looks Like Progress

Most AI failures are not caused by the technology itself, but by a lack of alignment between tools and business needs [1]. They happen when organizations jump into pilots without connecting them to real business needs. AI turns into a set of isolated efforts: a chatbot here, a copilot there. It feels productive, but it does not change how the business actually runs.

This trap starts quietly, fueled by good intentions. Leaders feel pressure to move fast and hope that doing something will create momentum. The first demo looks impressive, and a small pilot produces a quick win. It feels like progress. But speed without direction is just noise.

I learned this early in my career at a global company with 70,000 employees. We were struggling with a new applicant tracking system, and leadership pushed for AI features, convinced that smart automation would fix adoption problems. But the real issue was not the software. It was the people and the process. The system did not match the teams who used it. International groups ran into language barriers, applicants abandoned the process, and the marketing team that controlled the budget had never been consulted. The leadership chose the technology solution before properly defining the problem.

Adding AI on top of that did not fix it, it made it louder.

The same pattern shows up everywhere. When AI work is not tied to a clear business priority, it tends to drift. Small pilots solve interesting problems instead of important ones, and leaders start to mistake activity for impact. That illusion is dangerous because it convinces everyone that the strategy is working while the business slides back into the same problems, now with extra tools on top.

The data backs this up. The 2025 Stanford AI Index Report highlights a stark paradox: while 78% of organizations are using AI, only 1% have reached true maturity [1]. We are doing more with AI, but we are not improving. That gap between activity and impact is where most SMBs get stuck.

Leaders often start by asking:

"What tool should we use?"

When they should be asking:

"Where should we focus to move the business, not the technology?"

The truth is simple

AI does not transform a business on its own. It creates change only when it is tied to a clear outcome, a real priority, and a workflow people already use. Most small and midsize businesses do not need complex models or massive investments. They need focus: What should we tackle first? What should we avoid? What can we realistically support with our people and our data? And which outcome matters most right now?

When companies start with alignment instead of tools, everything shifts. Teams understand the purpose behind the work, so pilots target real operational problems. Early wins build confidence instead of confusion, and training feels relevant. Tools slip into daily routines rather than sitting unused on the shelf. Most importantly, the business starts to see repeatable, measurable impact.

That is why this chapter starts with the hard truth. If leaders understand where AI efforts go wrong, they can avoid wasted time, pilot graveyards, and the frustration that has slowed so many organizations. With alignment in place, the opportunity gets much bigger. The rest of this book is about how to get there, one clear and practical step at a time.

Chapter 2.2

Why an AI Strategy Matters

After looking at the common traps, it is time to shift the tone. A good AI strategy does not make work harder, it makes work clearer. It gives leaders and teams a shared direction so they do not waste time on tools that look exciting but lead nowhere. It is the bridge between experimenting with AI and actually benefiting from it.

At its core, an AI strategy does four things.

First, it brings clarity. AI is a crowded space, and without structure, teams do not know where to begin. Strategy defines the outcome that matters most and focuses effort on how AI supports it.

Second, it creates alignment. Without alignment, teams run disconnected experiments in parallel: an automation here, a platform evaluation there, a chatbot on the side. A strategy unites goals, priorities, and a shared definition of success.

Third, it protects resources. AI is accessible, but it still costs time, attention, and learning. Strategy forces focus on one or two use cases that will move the business, rather than spreading effort across too many pilots.

Finally, it accelerates value. Strategy is not just about order, it is about ROI. MIT Sloan research suggests the biggest jump in returns happens when organizations move from pilots to industrialized systems (the "Activate" phase) [2].

When teams have clarity and alignment, early wins come faster. People stop asking whether this works and start asking what's next. Strategy does not slow the organization down, it creates the conditions to move faster with less risk.

An AI strategy is more than a plan. It is how leaders steer through rapid change with confidence instead of impulse. It sets the rules of engagement so innovation happens in the right places, at the right pace, and with the right guardrails.

For small and midsize businesses, this kind of clarity matters. Time, budgets, and technical capacity are limited. Strategy turns those limits into design constraints, not barriers, so even small teams can use AI in meaningful ways.

A good strategy keeps the work doable. It shows where to begin, what matters most, and what can wait, so the organization moves with purpose instead of jumping from tool to tool. Below is a concise summary of what an AI strategy provides and how this book will help you build one.

What does this change day-to-day

A strategy changes the weekly reality of AI work. It gives teams a shared "yes" and "not yet," so you stop chasing scattered ideas and start finishing the few that matter.

Practically, it tends to show up like this:

- Fewer parallel pilots and side experiments competing for attention
- Clearer priorities and faster decisions about what to do next
- Less tool sprawl, because new tools are chosen for a defined workflow, not curiosity

- More realistic plans, because the work is sized to your people, your data, and your current capacity
- Lower risk without slowing down, thanks to clear guardrails for privacy, quality, and responsible use

Most importantly, it helps you build a repeatable process for choosing, launching, and scaling initiatives, so you don't reinvent the process every time.

Chapter 2.3

Your Reality Today

Every organization wants to move fast with AI, but nobody starts from the same place. Before you pick a tool or launch a project, you need an honest picture of your reality today. Your starting conditions shape what you can take on, how quickly you can move, and which early wins will build momentum instead of frustration.

This is not a judgment of your business. It is a map. When leaders skip this step, they end up choosing projects that are too complex or too disconnected from daily work. When they start with a clear map, they make decisions that create traction right away.

Four parts of your reality matter most: culture, capacity, data, and technical skill. Together, they determine what kinds of AI projects your organization can actually support right now.

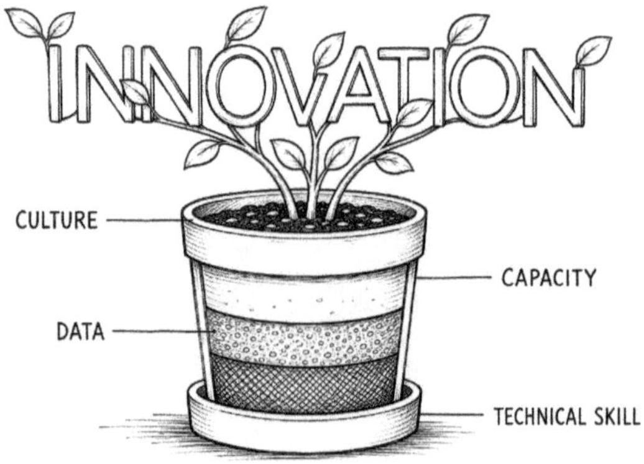

Culture: The environment for change

Culture shapes how people react to new tools, new workflows, and new expectations. It determines whether AI feels like support or a threat.

When people feel safe to try, learn, and make small mistakes, momentum builds fast. When fear or uncertainty is high, progress slows, and teams wait for perfect instructions.

Culture is your ceiling. Harvard's Amy Edmondson has shown that psychological safety (the ability to fail without punishment) is a strong predictor of innovation. If people feel they must hide mistakes, AI work will stall. [3]

Culture does not need to be perfect. It needs to be understood so leaders can introduce AI in a way that feels safe and aligned with the business.

Capacity: The time and attention you can allocate

AI takes energy upfront. Even simple use cases require time to explore, test, refine, and fit into daily work.

Capacity is bandwidth. Are teams constantly firefighting, or do they have breathing room? Can one person champion a small automation? Can a manager review a workflow? Do people have space to learn something new?

When capacity is low, start small with personal productivity wins or one targeted workflow improvement. When capacity is higher, you can take on more integrated projects with multiple stakeholders.

Data: The reality of what you have

Useful AI depends on the data you have today, not the data you hope to build later.

Some businesses have consistent, accessible information in core systems. Others rely on spreadsheets, manual entry, or processes that vary by person. What matters is knowing how structured your data is right now.

With clean, organized data, you can pursue analytics and integrated use cases. With messy or limited data, you will usually get better results by starting with workflow automation and productivity tools that do not depend on complex data pipelines. Early wins can improve data hygiene over time.

Technical skill: What your team can support

Technical skill is not about having data scientists. It is about what your team can maintain.

If nobody can support a custom solution, it will not last. If the team is comfortable with basic tools and low-code platforms, workflows can evolve quickly. If you have stronger technical capabilities, you can explore integrations and more advanced features when the timing is right.

Clarity here helps you avoid fragile solutions and choose projects your people can own, support, and improve.

Putting your starting conditions together

These four conditions are the foundation for choosing your first AI move. They help you separate what is possible from what is realistic. They prevent you from picking projects that are too big or too fragmented, and they often reveal strengths you may be underestimating.

When leaders understand the starting conditions, they introduce AI in a way that fits the business, rather than forcing the business to fit the technology. That creates early wins, reduces resistance, and builds trust.

The next chapter turns these starting conditions into action. It introduces a simple strategy framework to help you assess, audit, align, and activate AI in a structured, practical way.

AI Readiness Self-Assessment: The 60-Second Readiness Check

Chapter 2.3 explained your starting conditions. Use the radar in Figure 2.4 to map them across four dimensions.

This is not a test or a maturity score. It is a snapshot. Answer based on how things work today, not how you want them to work.

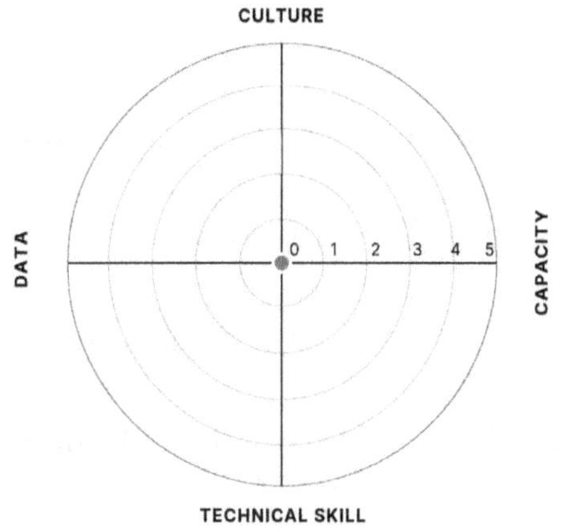

Figure 2.1. AI Readiness Self-Assessment radar

1. Culture

What it measures: Willingness to change, experiment, and learn.

Quick check:

- Do people feel safe trying new tools without fear of punishment?
- Are teams open to changing how work gets done?

- Do leaders encourage learning and experimentation?
- Are mistakes discussed openly rather than hidden?

Score (0-5): 0 = people hide mistakes · 5 = people share to learn

2. Capacity

What it measures: Time, focus, and clear ownership.

Quick check:

- Do teams have time to explore or improve workflows?
- Is there clear ownership for new initiatives?
- Are priorities stable enough to support follow-through?
- Can small improvements be tested without disruption?

Score (0-5): 0 = overwhelmed, no time · 5 = room to test and adjust

3. Data

What it measures: Structure, accessibility, and consistency.

Quick check:

- Is core data consistent across systems?
- Can information be accessed without heavy manual cleanup?
- Are processes documented or repeatable?

- Do teams trust the data they use daily?

Score (0-5): 0 = messy, trapped data · 5 = clean and accessible

4. Technical skill

What it measures: Ability to support and maintain solutions.

Quick check:

- Can your team maintain tools once they are deployed?
- Is basic automation or integration understood internally?
- Are vendors used intentionally rather than out of necessity?
- Can small changes be made without breaking systems?

Score (0–5): 0 = relies entirely on IT/vendors · 5 = can support in the workflow

How to score:

Pick one final score (0–5) for each dimension.

- Do not average question scores
- Do not score future goals
- Do not compare yourself to others

The shape matters more than the number.

How to read your shape:

Plot your four scores from Low to High and connect the points. Your outline shows what your organization can realistically support right now and where to start.

How to use this tool:

- Answer the diagnostic questions honestly to guide your thinking.
- Select a score (0-5) for each dimension based on where you are *today.*
- Observe your "Readiness Shape" generated on the wheel.

Two contrasting profiles illustrate the range. **Figure 2.2** shows a small, uneven shape typical of early-stage organizations. **Figure 2.3** shows a large, balanced shape typical of AI-ready organizations.

Example 1:

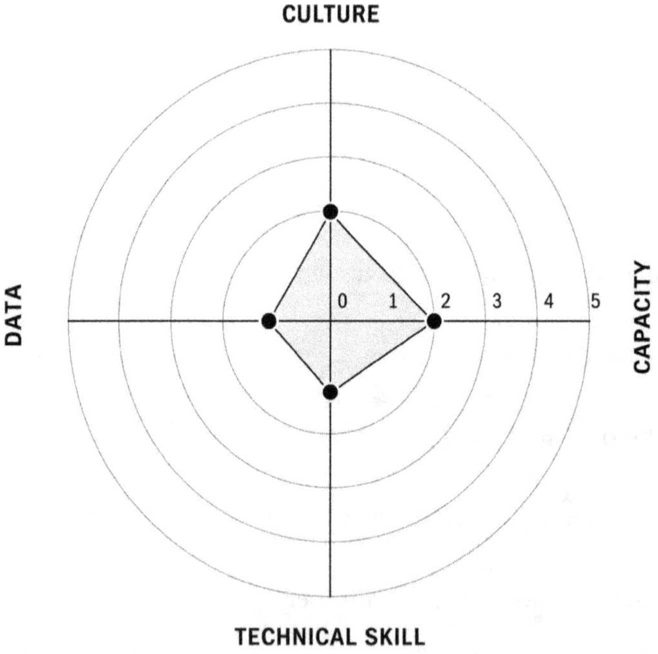

Figure 2.2. Example 1, Self-Assessment Radar - Smaller Shape

ANALYSIS:

When Your Shape Is Small

- Your current readiness for AI is limited.
- Avoid complex pilots or expensive enterprise software for now.
- Focus on individual productivity tools such as ChatGPT or Copilot.
- Use this stage to build confidence, basic fluency, and time savings.
- The goal is momentum, not scale.

Example 2:

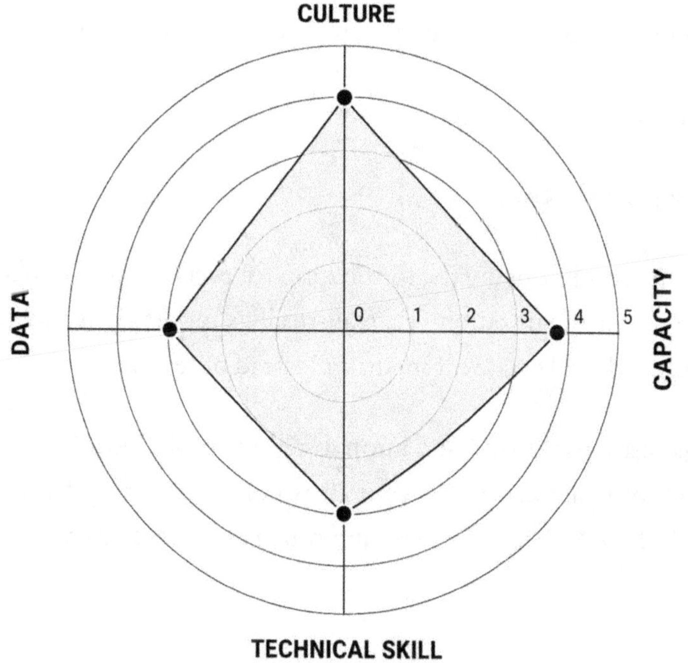

Figure 2.3. Example 2, Self-Assessment Radar - Larger Shape

ANALYSIS:

Diagnostic Result — Large Shape

- Your organization shows strong AI readiness.
- Core capabilities and data foundations are in place.
- Focus on scaling with confidence, not experimentation.
- Explore system integrations and multi-step workflows.
- Begin deploying customer-facing AI solutions with clear ownership.

The Golden Rule

Never choose a project that is bigger than your readiness shape.

Grow your shape first, then grow your ambition.

Bringing It All Together

Your starting conditions are the foundation for everything that follows. They show what your organization can realistically support right now, and what kind of AI work will create momentum instead of resistance.

Most organizations skip this and jump straight to tools. By doing the readiness check first, you gain clarity on your constraints and strengths. That makes your first "AI win" easier to choose and much easier to deliver.

This is the start of a real strategy.

The next step is turning that reality into a practical plan. In Chapter 2.4, I introduce the 4A AI **Strategy Framework,** a simple process for moving from assessment to action and building capability over time.

Chapter 2.4
The 4A AI Strategy Framework

The **4A** AI Strategy Framework **(Assess, Audit, Align, Activate)** is built for small and midsize businesses. It helps you identify the friction that matters, understand what you already have, choose a realistic approach, and turn it into a pilot you can ship.

You may already use other planning frameworks or maturity models. 4A does not compete with those. It fits on top of what you already do and adds structure without adding complexity. If you do not have a framework, it gives you a straightforward place to start.

At its simplest, 4A is a step-by-step process for turning clarity into a concrete plan. It is light enough for a 10-person business and still useful at 500.

Start with a team workshop

You do not need consultants or a retreat. Start with a 30-60 minute working session:

1. Bring the right people together.
2. Use the readiness shape from Chapter 2.3.
3. Have everyone mark where they think the organization stands.
4. Compare shapes and discuss the gaps.
5. Use the patterns to begin the 4A steps.

This creates alignment fast and gives you a shared starting point.

The Framework

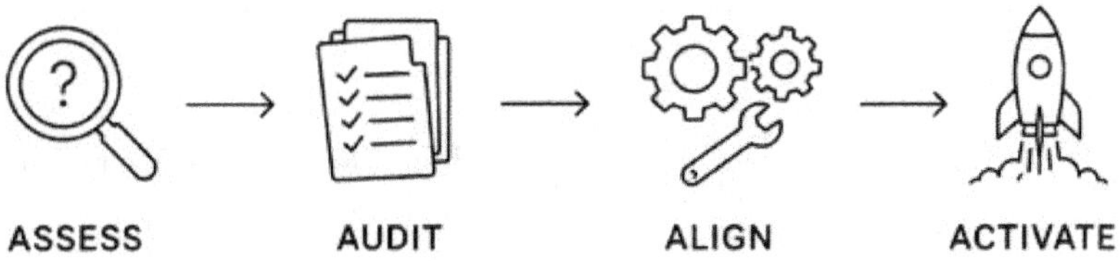

Figure 2.4. The Four-Step Framework: Assess, Audit, Align, Activate.

I first created this framework in 2011 while consulting on digital operations. It has four steps, shown in Figure 2.4: Assess, Audit, Align, and Activate. The sequence has stayed the same because it works: the steps help teams move from uncertainty to clarity, and from ideas to results

If "framework" sounds abstract, do not overthink it. It is simply a way to solve the problem in the right order, stay focused, and avoid the chaos of trying too many things at once.

We will make this concrete with an example: John, a small business owner with 14 employees. John represents many leaders who want to use AI, but do not have a technical team and cannot afford major disruption. You will see how each step works and how to apply it in your own organization.

Now, let us walk through the four steps:

Assess → Audit → Align → Activate

Step	What It Means	What You Do	Example (John's Company)
Assess	Identify the biggest friction point	Choose 1 priority to focus on	John chooses "manual customer updates"
Audit	Understand what exists today	Map the process, data, and tools	John discovers data is already in his CRM system
Align	Match the project to your reality	Check culture, capacity, data, and skill	High culture, low skill → simple automation with basic AI chatbots, for example
Activate	Build a small pilot	Start small, measure impact	Use AI to manage phone calls, chatbots, and social messaging while qualifying sales leads from general inquiries and support requests.

1. Assess: Identify What Matters Most

Small businesses cannot solve everything at once. Assess is about choosing the one problem that matters most right now, and matching it to what your organization can realistically support. Use your AI Readiness Shape from Chapter 2.3 as a guardrail here. The goal is focus, not ambition.

An AI strategy does not begin with tools or features. It begins with a clear decision about what deserves attention today. When everything feels important, nothing moves forward.

Choosing your lane

Your readiness stage shapes the kind of first project that will actually work.

If you are in the **early stage,** start with personal productivity. Look for wins that save an individual about 30 minutes a day, such as drafting emails or summarizing meetings.

If you are in the **friction stage**, aim at one bottleneck in a real workflow. The value comes from reducing repetition and errors, not from sophistication. Think invoice entry, lead capture, scheduling, or customer updates.

If you are in the **advanced stage,** you can take on system integration and analytics because the data foundation is already stronger.

A simple rule helps: do not pick a project "above" your readiness shape. Even good ideas stall when the organization cannot support them yet.

To pick the right starting point, ask one question as a team: **What slows us down or frustrates us the most every week?** In most SMBs, the answers are familiar: slow customer communication, manual rework, scheduling friction, late invoices, inventory errors, or slow onboarding.

Relief is a signal. The problem that causes the most irritation is often the best place to start.

Assess checklist

You are ready to move on when the team can name one focus area in plain language, agree on why it matters, and describe what "better" looks like.

John's example

John's customers constantly call for job status updates. Office staff spend hours rewriting technician notes into emails, customers reply with more questions, and messages get missed. The friction is obvious, so the priority becomes simple: **fix customer communication.**

2. Audit: Understand What You Have (and Don't Have)

Audit sounds technical, but it is not. It is a practical inventory: **what do we already have that AI can use?** Most of the time, the fastest path is not a new tool. It is using existing tools and data more deliberately.

To audit quickly, look at three things.

Your process

Start with what actually happened this week. Where does the work get stuck? Who touches it along the way? John walks a single customer update from the shop floor to the customer. The same note gets rewritten three times, re-sent by hand, and messages get lost.

Your data

Next, ask where the information lives and how reliable it is. In John's case, job updates already exist in the job management system. The problem is not missing data, it is how the team uses it.

Your tools

Finally, review the software you already pay for and what features are turned off. John finds basic automation built into the current system that nobody has enabled.

You do not need a data scientist. You do not need perfect data. You need a clear picture of the process, the inputs, and what is already available.

3. Align: Bring People and Priorities Together

Align is where a good idea becomes a realistic plan. The goal is to make sure the work fits the team that has to run it and the reality you are operating in today. For SMBs, this step matters more than picking the "best" tool.

A simple way to align is to answer three questions together:

1. Does this solve a real problem we actually feel?

2. Does it align with our current reality (culture, capacity, data, skills)?

3. Will it require meaningful behavior change, and are we ready to support that?

For John, the picture is clear: culture is high, capacity is medium, data is solid, and technical skill is low. That rules out heavy integration for now and points to a lightweight automation inside an existing tool.

Alignment is what keeps you from building something impressive that nobody adopts.

4. Activate: Start Small, Build Fast, Measure Clearly

Activate is where you turn the plan into a pilot. Keep it small, keep it low-risk, and tie it directly to the friction you chose in Assess.

John's pilot is simple: he enables automated customer update templates in the existing system, uses AI to summarize technician notes, and sends a weekly status email without having to rewrite everything by hand. He tracks two outcomes: hours saved and fewer customer calls.

The setup takes two hours. Within a week, the office staff gets five hours back, and customer calls drop by 40%.

Activation is not the end. It is the beginning of momentum. Treat the first pilot as a learning loop: ship something real, measure it, and improve it.

Putting the 4A AI Strategy Framework Together

The power of the 4A strategy lies in its simplicity. It provides a structure that small businesses can actually follow, even without an IT team, data experts, long planning cycles, or a technical background.

This is a strategy process that fits the real world. It works for companies with five employees or five hundred.

Now that you have a clear method for building a simple and practical AI strategy, the next chapter will help you choose your first AI move based on your readiness and the opportunities in front of you.

Chapter 2.5

Applying the 4A AI Strategy to a Real Business Scenario

Imagine a residential inspection company called Novexa Property Inspection Services. It is an 11-person team. Their onsite work was efficient, but turning photos and spoken notes into a polished client report took 48-72 hours. Admin rewrote everything manually, inspectors reviewed late at night, and realtors grew frustrated with delays. Without a development team or budget for custom tools, Sarah, the owner, brought the team together for a 4A AI Strategy Framework Workshop to determine whether AI could meaningfully reduce this reporting bottleneck.

Novexa's 4A Strategy Summary After the Workshop

The workshop produced a short internal strategy document built around a single bottleneck: reporting was slow, admin time was getting crushed by rewriting, and clients and realtors were feeling the delay. The team did not need a grand transformation. They needed a first win that reduced turnaround time without adding a custom build.

The workshop's output was not a slide deck. It was a simple decision and a short plan.

Novexa chose one priority: reduce report turnaround time. The team mapped the current workflow (rewrite → edit → review → PDF), confirmed the inputs were already strong (photos, categories, CRM fields), and spotted the real bottleneck: transcription and repetitive rewriting. Given their readiness and capacity, they avoided a heavy build and designed a lightweight 30-day pilot they could actually ship.

That is what it looks like to turn “we should use AI” into a strategy: one clear focus, a grounded view of what exists, a realistic plan for the team, and a measurable pilot.

Turning Strategy Into Action

Instead of buying a new platform, the team built a lightweight workflow with what they already had: CRM + transcription + Integration Tool + an AI step.

Novexa’s AI-Enabled Workflow

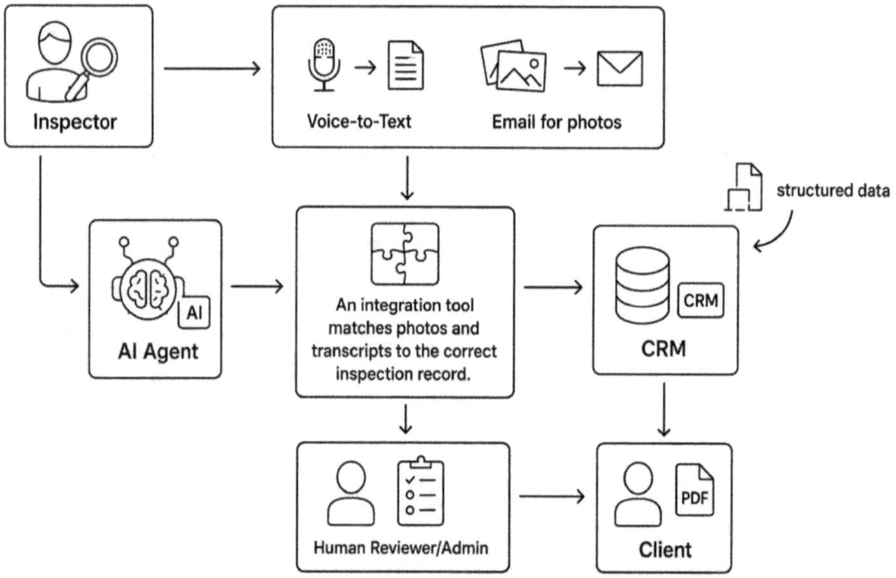

Figure 2.5. Automated inspection workflow, from voice notes to client delivery.

1. Inspectors record voice notes, and a transcription app converts them to text.

2. Photos are emailed to a dedicated inbox (e.g., “INSPECTION_REPORT – [Address]”).

3. An integration tool matches photos and transcripts to the correct inspection record.

4. An AI step converts the transcript into structured fields (category, severity, recommendations, photo references).

5. Your CRM generates a draft report, the admin reviews, the inspector approves, and the PDF is sent.

The workflow shown in Figure 2.5 dropped turnaround time from 48 to 72 hours to same-day delivery, and the administrative workload decreased significantly.

What Small Businesses Can Learn From Novexa

Novexa is a good reminder that the first AI win is usually not a flashy model. It is a practical workflow fix. Start where the friction is highest, use the tools you already pay for, and design the pilot around what your team can support. AI can remove repetitive work, but people still own judgment and final approval. One solid pilot makes the second one easier to sell and faster to run.

From Strategy to Execution

A single pilot is the beginning, not the end. Once Novexa saw a measurable impact, they needed a roadmap to guide what comes next. A roadmap sets milestones, scales AI into new processes sustainably, and ensures early wins turn into long-term capability.

In the next chapter, we shift from a single example to a universal approach. You will learn how to turn your strategy into a 90-day sprint and a 12-month AI roadmap that any organization can use to move from an early pilot to sustained, repeatable progress.

Chapter 2.6

The Roadmap and the 90-Day Strategy Sprint

A strategy becomes valuable only when it turns into consistent action. Most small and midsize businesses do not need a multi-year transformation plan. They need a practical starting point, a clear sequence of steps, and a realistic timeline that aligns with the pace of their people and resources.

The purpose of this chapter is simple. It gives you a structure to move from a strategy on paper to a working AI initiative that produces measurable results. The process includes two parts:

1. A **90-Day Strategy Sprint** to deliver your first real win.

2. A **12-Month AI Roadmap** to extend that momentum into broader capability.

This chapter is intentionally generic. The roadmap is designed so any business, regardless of industry or size, can pick it up and apply it immediately.

Why a Roadmap Matters

A roadmap does not predict the future, it prevents you from improvising every week.

When teams skip this step, they either try too much too soon or stall after the first experiment. A simple roadmap does the opposite: it protects focus, sequences the work, and gives people permission to say "not yet" to anything that does not serve the current sprint. With a roadmap, it is easier to define the goal for each quarter and measure progress.

The 90-Day Strategy Sprint

The goal of the sprint is simple:

Deliver one meaningful AI win in 90 days.

This creates momentum, builds confidence, reduces fear, and gives your team a template for future projects.

Here is the structure.

Phase 1: Preparation (Weeks 1-2)

Focus: clarity and boundaries.

Start by confirming the priority use case. Refer back to your 4A Strategy Summary and choose the workflow that creates the most friction today and offers the highest payoff if improved. Do not overthink it-pick the pain point people complain about every week.

Next, define success metrics so you will know whether the sprint worked (for example: hours saved, fewer errors, faster turnaround, fewer customer follow-ups). Then assign clear roles: a Project Owner to drive decisions, a Technical Implementer to build, and a Reviewer to validate quality and risk.

Finally, set boundaries. Decide explicitly what you will not attempt in this sprint. Small wins succeed because they stay small.

Phase 2: Build and Test (Weeks 3-6)

Focus: experimentation and iteration.

Build a v1 of the workflow-the "wiring" that makes the process run end-to-end. This usually includes prompt design, basic automation, and mapping fields between the systems you already use (CRM, ticketing, spreadsheets, email, shared drives, etc.).

Test the workflow on a small sample before expanding it. Use 5-10 real cases, not hypotheticals. Then iterate: tighten prompts, add validation where the AI output could create rework or risk, and simplify any step that feels clunky.

Phase 3: Deploy and Measure (Weeks 7-12)

Focus: adoption and proof.

Roll the workflow into real work. Adoption is the goal here-use it on live tasks, in the normal weekly rhythm, with the human review step clearly defined.

Track progress weekly against the metrics you set in Phase 1. Pay close attention to where people make manual edits. Those edits are your best signal for quality gaps and training needs. Collect feedback from the people in the loop: what is confusing, what is working, and what still feels slower than the old way.

Close the sprint with a one-page review summarizing impact, lessons learned, and the recommendation for what happens next. That page becomes the "permission slip" for the next sprint.

Keep in mind (to prevent sprint failure): In every phase, protect time, keep scope narrow, and make adoption measurable. If any of these slip (unclear scope, no real bandwidth, missing inputs, or low trust/adoption), pause and simplify before moving forward.

If you run into resistance or stalls mid-sprint, the patterns below are the most common and the fastest to fix.

The Sprint Friction Radar

Why sprints stall and how to unstick them

Even with a roadmap, human issues will pop up. Use this radar to identify friction early.

If you see this...	It usually means...	The fix
"We don't have time to test this."	The project is seen as extra work rather than a replacement for work.	Redirect Capacity: Don't add the AI project to the to-do list. Pause a non-essential recurring meeting or low-impact task for 2 weeks to create a focused 'test budget' for the pilot.
"The AI output is too generic."	Your prompts lack context or examples.	Deepen Prompt Inputs: Stop asking for broad results. Provide the AI with 3 'gold standard' examples of the output you want. Define specific target audience, tone, and constraints before running the prompt.
"We are arguing about features."	Scope creep has set in.	Return to the Core: Stop feature-creep. Strip the project back to the 'Minimum Viable Workflow' that solves the single biggest bottleneck. If it doesn't solve the primary issue immediately, cut it.
"It works, but nobody trusts it."	Fear of replacement or lack of transparency.	Make Human Review Explicit: Move the 'Human Approval' step to the center of the workflow. Explicitly define the human as the final editor and the AI as the assistant, ensuring the human's contribution is valued and required.

Budgeting for the 90-Day Sprint

You do not need a large budget to start. In fact, throwing money at the problem before you have a process is a recipe for waste. But you do need to align your spending with your stage.

Once the sprint is unstuck, the next question is usually cost. Here is how to budget without wasting money.

Most successful sprints operate in this range:

Your stage	Est. monthly cost	Best ROI opportunity	What to avoid (the money pit)
Level 1: Enablement *(Personal productivity)*	**$25–$60** (per user)	**Time savings:** drafting content, summarizing emails, fixing code.	**Consultants.** Do not hire anyone. This is a DIY stage.
Level 2: Automation *(Process fixing)*	**$100–$500** (tool subscriptions)	**Error reduction:** automating data entry, invoices, and scheduling.	**Avoid custom software development.** Stick to off-the-shelf automation tools.
Level 3: Integration (*Data connection)*	**$2,000+** (one-time setup)	**Decision quality:** unified dashboards and real-time reporting.	**Cheap fixes**. Do not try to DIY an integration. Hiring a pro here saves money later.

Essential tools checklist

- **AI platform**
- **Automation Tool**
- **Existing stack:** your current CRM or operational system

Human resources

The real cost is focus. You need to budget time:

- **Project Owner:** 2–3 hours per week
- **Technical Implementer:** 5–8 hours per week
- **Reviewer:** 1 hour per week

Note: If you lack internal technical skills, a freelancer can set up most SMB-level automations for a one-time fee of $600 to $2,500. Use an LLM to write the specifications for them to keep costs down.

The 12-Month AI Roadmap

Once you complete your first sprint, the next step is expanding AI across the business. The goal is not to automate everything. The goal is to build capability quarter by quarter.

Here is a sample roadmap that fits most SMBs.

Quarter 1: Build the first win

Complete the 90-day sprint, document the workflow so someone else can run it, and make the impact visible. End the quarter by identifying the next two candidates for improvement based on real friction you observed.

Outcome: one working AI-assisted process and a confident team.

Quarter 2: Strengthen foundations

Make the work repeatable. Standardize naming, file storage, and internal data structure. Turn your best prompts and automations into reusable assets. Train at least two additional team members to maintain AI workflows, and add a simple validation checklist to reduce errors.

Outcome: a stronger internal foundation that is easier to support.

Quarter 3: Expand to a second workflow

Choose one additional high-friction process and implement it using the same sprint structure as before. Good candidates usually include customer support and client communication, scheduling, reporting, documentation, internal analytics, lead qualification, content planning, or ad creative production.

Outcome: AI capability expands into a second business area.

Quarter 4: Integrate and optimize

Refine automations based on real usage, expand training when it makes sense, introduce dashboards to track performance, and evaluate which use cases now justify deeper integration.

Outcome: the organization shifts from experimenting to operationalizing.

How to use this roadmap

Treat this roadmap like guardrails. If your team is slammed, slow down. If you have a rare window of focus, move faster. The only rule is that the work has to fit into real life and still make sense, because the goal is **momentum you can keep.**

A simple way to stay on track is to end each quarter with three wins: one workflow that works better than before, one person who leveled up, and one part of your system (documentation, data, automation, or reporting) that is stronger than before.

Do that consistently, and a single pilot turns into a capability you can repeat. As the team gains experience, execution gets smoother, confidence grows, and the wins start to compound.

Template

Your 12-Month AI Roadmap

A roadmap you can adapt to your business, with prompts and checklists for each quarter.

Scan or visit **designated.ai/books/practical-strategy**

Chapter 2.7
Strategy Red Flags

Even with a clear framework and a strong roadmap, many organizations still stumble when putting AI into practice. The good news is that most failures follow predictable patterns. Once you recognize them, you can avoid wasted time, unnecessary spending, and stalled progress.

This chapter covers the most common early red flags and the practical alternatives that keep your strategy on track.

Red Flag 1: Starting with tools instead of outcomes

Many leaders begin by exploring tools. They buy licenses, test features, and experiment with prompts. That creates activity, but not progress. When the focus is on the tool rather than the outcome, results rarely scale.

Tools do not solve business problems by themselves. They become valuable only when applied to a specific workflow that needs improvement.

This red flag also shows up when organizations copy competitor initiatives or chase advanced use cases just because others are doing it. "We need what they have" is usually a sign the team has lost contact with its real constraints and priorities.

Healthy alternative: Start with the problem and define the outcome first. Then let the outcome determine the tool path. Ask: *What process needs improvement?* Not: Which tool should we buy?

For creative or public research, everyday AI tools may be sufficient. For private-company data, built-in AI within existing platforms may be a better fit. The choice should follow the outcome, not the other way around.

Red Flag 2: Skipping the foundations

Culture, capacity, data, and skill are the foundation of AI readiness. When leaders skip an honest assessment, they often choose projects the organization cannot support.

The warning signs are predictable: a busy team is handed a complex initiative, data is inconsistent but leadership pushes for predictive analytics, or technical skill is limited and someone proposes a custom solution.

In many cases, the issue is not that the organization lacks data. It is that the data is misunderstood, fragmented, or not ready for advanced use cases.

Healthy alternative: Match the size of the project to the size of your readiness. If your shape is small, your first project should be small. Start with workflows that use the data you already have, and let better data practices emerge over time.

Red Flag 3: Choosing workflows that require major behavior change

If a pilot forces employees to fundamentally change how they work, it will struggle unless adoption is heavily supported. New documentation habits, new communication steps, or new review processes add friction that slows progress.

Behavior change takes time. When too much change is required too early, resistance increases and usage drops, even if the technology itself works.

Healthy alternative: Start with solutions that fit naturally into existing habits or run quietly in the background. Save larger behavior shifts for later stages, once trust and confidence are established.

Red Flag 4: Taking on projects that are too big

Ambition is healthy, but large projects overwhelm small teams. Attempts to automate an entire department, migrate all systems at once, or redesign every workflow often lead to delays and frustration.

Big projects fail quietly. They consume time, stall decision-making, and drain momentum before delivering visible value.

Healthy alternative: Use the 90-day strategy sprint. If a project cannot be completed in 90 days, it is too large for a first win. Smaller wins build confidence and create the foundation for more ambitious work later.

Red Flag 5: No human review layer

Early AI workflows need verification. When teams skip review, errors accumulate, trust erodes, and rework increases. The absence of a review layer also creates fear, especially in customer-facing or decision-critical processes.

Healthy alternative: Adopt the principle of accountability. The AI drafts, the human edits. Include a simple human review before anything is sent or acted on. Over time, the layer may become lighter, but it should never disappear entirely.

Red Flag 6: Treating AI as a one-time project instead of a capability

Some organizations complete a single pilot and assume the work is finished. AI is not a one-time initiative. It is a capability that grows through repeated use, refinement, and learning. [4]

The real economic value comes from sustained productivity improvement. It is built gradually, not unlocked all at once. [5]

Healthy alternative: Adopt a quarterly rhythm. Each quarter should deliver one improvement, one new skill, one strengthened system, and one step forward in maturity. This turns early wins into durable capability.

Bringing it all together

These red flags appear in organizations of every size and industry. They are not signs of failure. They are signals to slow down, re-align, and choose projects that match your readiness.

When you avoid these pitfalls, AI efforts become practical, stable, and repeatable. Risk drops, trust grows, and teams gain confidence to move at a pace they can sustain.

You now have the foundation for building an AI strategy:

- Realistic starting conditions
- A practical framework
- A clear first pilot
- A roadmap for growth
- Guardrails that protect progress

The next stage is execution at scale. With these principles in place, you are ready to move from early wins to long-term capability.

Chapter 2.8

Conclusion: The Path Forward

Every organization starts in a different place, but the path forward with AI follows the same logic. Meaningful progress does not come from big platforms, big budgets, or ambitious technology plans. It comes from clarity, focus, and the discipline to improve one important workflow at a time.

An effective AI strategy starts with a clear view of your reality. Choose a problem that truly matters, then align the work to what your team can actually support. When you execute that first focused improvement, you create momentum that cannot be bought or outsourced. Confidence grows, time comes back, and the organization develops a repeatable way of working.

The goal is not to "launch an AI initiative." The goal is to deliver one concrete result that proves AI can help the team work better, faster, and with less effort. Once that first win is in place, the next step becomes clearer, and progress becomes easier to sustain.

To help you put this into motion, this book includes practical tools you can use immediately. The templates and worksheets in the resource section are designed to help you move from ideas to execution with focus and clarity.

If you can ship one real win, you can build the capability to ship the next.

References

1. Maslej N. Artificial Intelligence Index Report 2025. Artif Intell. 2025.

2. Grow Enterprise AI Maturity for Bottom-Line Impact | MIT CISR [Internet]. [cited 2026 Aug 22]. Available from: https://cisr.mit.edu/publication/2025_0801_EnterpriseAIMaturityUpdate_WoererSebastianWeillKaganer

3. Edmondson AC. The fearless organization: creating psychological safety in the workplace for learning, innovation, and growth. Wiley; 2018.

4. McKinsey & Company. The state of AI: agents, innovation, and transformation [Internet]. 2025 [cited 2026 Aug 22]. Available from: https://www.mckinsey.com/capabilities/quantumblack/our-insights/the-state-of-ai

5. Chui M, Hazan E, Roberts R, Singla A, Smaje K, Sukharevsky A, et al. The economic potential of generative AI: the next productivity frontier [Internet]. McKinsey & Company; 2023 [cited 2026 Aug 22]. Available from: https://www.mckinsey.com/capabilities/mckinsey-digital/our-insights/the-economic-potential-of-generative-ai-the-next-productivity-frontier

Part 3

MAKING IT WORK

Chapter 3.1
Making AI Part of Your Culture

Your business rolled out a new AI tool a month ago. Same announcement to everyone, same access, same nudge to give it a try. Look at your team now. One person uses it every single day and wonders how she ever worked without it. Two others on the same team never really bothered, and if you're honest, you're not sure they even signed in.

Sound familiar? The tool worked fine. What wasn't ready was the culture around them, the thing that has to carry any new tool. It's worth being clear about what culture means, because it's easy to misname. Culture is not the poster on the wall or the values page on the website. It's how work actually happens around here, how people behave when nobody's watching and no boss is standing over them. That is what decides whether AI does anything for you, because a great idea is worth nothing until someone actually uses it. You can buy every tool on the market, and nothing changes until your team decides to pick it up.

Part 2 gave you the plan. You assessed where you stand, audited your process and your data, and mapped a roadmap for the months ahead. This chapter is where that plan meets the people who have to implement it. That work lands on you. Your team watches what you do more than they hear what you say, so you set the tone either way.

This chapter and the rest of Part 3 help you and your team actually put AI to work. Each chapter comes with worksheets, prompt packs, and checklists you can grab from the QR codes along the way.

Free companion toolkit

Get the Part 3 Worksheets

Prompt packs, checklists, and the Adoption Sheet, organized by chapter

Scan or visit **designated.ai/books/practical-strategy**

Fast Movers, Slow Movers, and You

Let's start with how this actually feels day to day, because there's a lot of anxiety around AI right now, and inside a business it usually shows up as frustration.

Something feels off about the pace. Either it's moving too slowly, and you're watching people drag their feet while the opportunity sits there, or it's moving too fast, and someone on the team is charging ahead, using tools nobody vetted, making a mess you'll have to clean up. Whichever it is, you've probably already got a name in your head for who's responsible.

That reaction is completely human. The trouble is that the name in your head is usually the wrong thing to focus on, and stewing over it changes nothing. Before we talk about what to do, it helps to understand where the frustration actually comes from.

Here's the thing most people never learned. Whenever anything new shows up, a tool, a process, a way of working, people don't take to it with the same ease or at the same speed. A few jump on it immediately. A few more follow once they've watched the first group. Most sit in the middle and wait for proof. A few hold out until they have no other choice. This isn't a flaw in your team. It's how every group of humans has taken up every new thing, and researchers have watched it play out the same way for decades, across farming, medicine, technology, all of it.[1]

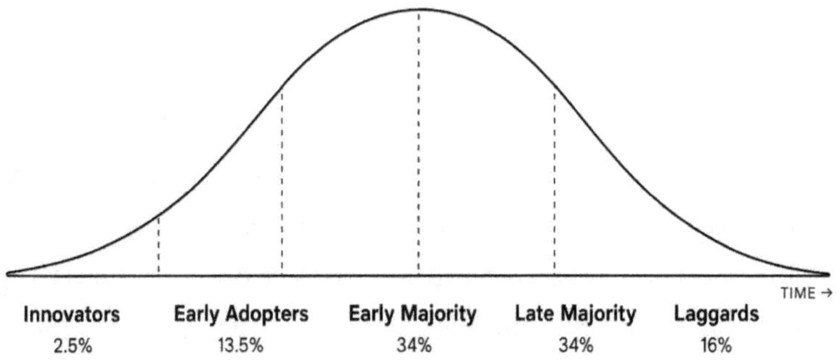

Figure 3.1. Rogers' Diffusion of Innovations. The five groups that appear whenever something new spreads through a population.

As Figure 3.1 shows, the same pattern plays out every time. Your fast movers aren't reckless, and your slow movers aren't lazy. They're standing in different places on the same curve, and both ends have their reasons. The fast mover is chasing what the tool might do for her. The slow mover has watched shiny promises come and go and is waiting to see if this one's real.

Here's the part that stings a little. You're on this curve too. Take me. I'm an early adopter of AI, genuinely excited, because I can see exactly where it earns its keep in a business. Put crypto in front of me and I'm the laggard at the back of the line. It doesn't interest me, I don't use it, and I've felt no pull to change that. Same person, opposite ends, depending entirely on whether the thing lands somewhere I care about [2] . When you catch yourself frustrated that someone on your team is dragging their feet, remember there's something out there you're dragging your feet on too, and you've got perfectly good reasons, same as they do.

That's what "and you" means in the title of this section. The frustration isn't really about them being wrong. It's the gap between where you are and where you think everyone should be. Drag people across that gap and you get resentment. Understand why they're standing where they're standing and you get movement.

Here's what gives me hope, and it's something we've seen again and again. Where someone stands isn't fixed. Change what's around a person, how safe it feels to try, what they see the people above them doing, whether the reason to bother is clear, and they move. The environment shifts the position. That's the whole reason this chapter exists, because the environment is yours to shape. The pattern has a shape, and once you've seen the shape you stop taking it personally.

Worksheet

Rate Your Culture

Score where your team stands.

Scan or visit **designated.ai/books/practical-strategy**

What People Are Protecting

What actually holds someone back? Once you look past the frustration, it's almost always the same thing.

Watch two people meet the same new tool. One sees what it might do and jumps straight in, gets a dozen bad answers and doesn't care, because the one good answer coming is all that matters. The other sees everything that could go wrong, the wasted time, the mistake in front of a client, and hangs back. It's someone protecting themselves from a failure they can already picture. One person is looking at what there is to win, the other at what there is to lose, and that's what decides who jumps.

What decides which way a person leans is where the tool lands. Take an estimator at a general contractor. His whole job is to look at a project, work out the materials, the labor, the timelines, and put a price in front of the customer that the company bids on. Get it wrong and you either lose the job or lose money building it, so over twenty years he's become the person everyone trusts to get that number right. That's his standing in the business. It also takes him the better part of a week per bid.

Give that estimator AI for routine emails to suppliers and he shrugs and uses it, no problem. Hand him one of the newer tools built for estimating, and something in his face closes. These tools don't just pull live pricing off the market. They act. They gather the data, chase down what's missing, and put together a draft estimate on their own, in a fraction of the time it takes him. He's still needed, though not for the same thing, and he knows it.

That is the part worth sitting with, because his unease is not paranoia. His role really is changing. The week of building an estimate from scratch is the thing he was best at, and a tool just did a rough version of it in an afternoon. What's left for him is judgment, checking the tool's work, catching what it got wrong, deciding what to trust. That may well be the most valuable part of the job. It's not the part he built his name on.

Worksheet

What Each Person Is Protecting

Discover what each person is known for.

Scan or visit **designated.ai/books/practical-strategy**

This is the pattern underneath most resistance. People don't resist AI. They resist AI taking over the part of the job they're known for, and the standing that comes with it. It isn't about age, and it isn't about how long they've been here. In the research, how threatened someone felt predicted resistance strongly, while age predicted nothing at all [3]. One study even found newer people more rattled by a tool than veterans with over a decade behind them, because the newcomers had less standing and more to lose [4]. What matters is whether the tool reaches into what makes them valued.

If you want to understand a specific person's resistance, ask why, and don't stop at the first answer. Ask the estimator and he'll tell you the tool makes mistakes. Push a little and it's that a bad quote costs real money. Push once more and you reach the truth, pricing jobs is his reputation, the thing people come to him for, and this tool feels like it's coming for it. The mistakes were only ever the surface. What he's really protecting is his standing.

Here's what that changes about how you lead it. Don't calm him by pretending nothing's different. He can see that it is, and denying it just tells him you're not paying attention. Name the change honestly, then show him where he fits in it. The judgment was always the hard part of his job. Now it's the whole job, and his twenty years are exactly what make the tool safe to rely on. That answers the fear he actually has. Telling him to trust the data answers one he doesn't.

Before any rollout, run this for each person the tool will touch. Name what they're known for, then ask whether the tool operates there. When it does, expect resistance, and don't read it as stubbornness. From where they're standing, it's the sensible reaction. We call this the Advantage Check, and there's a simple version on the resource page you can run in ten minutes.

Worksheet

The Advantage Check

A ten-minute worksheet to spot where each person's expertise overlaps with the tool, so you can lead the change before it lands.

Scan or visit **designated.ai/books/practical-strategy**

It also tells you why the big announcement never works. The worry underneath is about standing, and no speech touches standing. You can't talk someone out of a fear that the thing they're good at is about to matter less. You can only show them, over time, that it isn't true, or that the new version of their job is one they'd actually want. That's the work the rest of this chapter is about.

What a Good Culture Looks Like

You've seen what people protect and why they hold back. The question now is what you build in its place. What does an organization actually look like when its culture is ready to take on AI.

Start with the thing everything else rests on. People understand where the business is going. They can tell you the vision and the direction for the next couple of years. They know the core values too, the ones that guide how the work actually gets done. More than knowing all this, they act on it. Faced with a decision when no one is watching, they make the call the business would want. That is culture doing its real job, carrying the strategy into small daily choices no leader is there to see.

Here's why that comes first. AI is far easier to judge when the fundamentals of the business already work. When people know where the business is headed, they can look at a tool and see whether it moves the company that way or just adds motion. When the direction is fuzzy, every shiny tool looks equally plausible, and you end up chasing features instead of outcomes. AI doesn't fix a business that's unclear about where it's going. It just makes the lack of clarity more expensive. The stronger these fundamentals, the more use you'll get from these tools, which is the real reason culture comes before software.

On top of that foundation, a few practical things have to be true.

People feel safe, and they tell each other the truth. Someone can ask a naive question, admit a tool confuses them, or say an idea isn't working, without it costing them anything.[5] Feedback is honest, with none of the fake politeness that lets a bad approach limp along because nobody wanted to say so. When something fails, and a good half of what you try with AI will, the person who tried it and shared what they learned gets met with thanks, not blame. Punish a failed experiment once and the next one you'll never see, only the silence.

People are encouraged to push. Raising an idea, questioning the way it's always been done, suggesting a tool worth a look, all of it is welcomed rather than treated as overstepping. Each quarter people bring what they've found, the wins and the dead ends both, and the best of it gets pointed at the goals everyone already knows. You still decide what gets adopted. What you're building is a steady flow of ideas coming up from the floor, aimed in the same direction.

People trust that improving the work won't cost them their place. This is the fear sitting under everything in the last section, so answer it straight. Nobody can promise every role stays exactly as it is, because these tools will reshape some of them. What you can promise, and mean, is that a person who improves the work is worth more to you and not less, and that if AI reshapes or removes part of what someone does, you retrain them and deal with them fairly rather than discarding them. When people believe that, they stop guarding the old way and start improving it.

Companion Resource

Building a Culture That's Ready for AI

A short playbook for lifting each part of your culture.

Scan or visit **designated.ai/books/practical-strategy**

None of that is about the software. Get these right and something shifts. Your team stops bracing against new tools and starts reaching for them. The person who moves AI forward often isn't the one with the title. It's whoever's curious enough to try things and generous enough to show others what worked. In a smaller business that might be a frontline employee who started experimenting on their own time, not a manager and not anyone with "technology" in their job description.

A good culture lets that person lead without needing a title to do it, and the leader's part is to say out loud that helping the rest of the team is welcome, so it never reads as stepping out of their lane. That's what a ready culture buys you, a team that carries AI forward instead of one you have to push.

Culture is hard to pin down. You can't put it on a dashboard the way you can revenue or cycle time, and that's exactly why it gets neglected. A good leader feels its weight anyway. The teams that win have it, because your best people only do their best work in the right environment, and they'll leave to find one if they have to. I've seen both kinds of places, and the difference is night and day. Building it takes real commitment from a leader, and that's what the next chapter is about.

Chapter 3.2

What Your Team Needs to Succeed with AI

In the previous chapter, you saw why making AI part of your culture is essential for real progress. Now, let's look at what your team actually needs to succeed with AI in practice.

Lena was sharp, charismatic, and always in motion. She loved ideas the way some people love coffee, intense, frequent, and best served fresh. She'd fly out to an industry summit or tradeshow and return brimming with excitement.

One week showed the whole pattern.

Monday, she swept into the leadership meeting fresh off a flight. "I met someone on the plane. They've got AI that redesigns your whole onboarding flow. Ours is embarrassing next to it. We need this." The team set aside the feature they'd been building for a month and prepared to pivot.

By the Tuesday sales meeting, she had another one. An AI system that scored and routed every inbound lead on its own, worth rolling out now, while the pipeline review waited. "Game-changing," she said.

By Friday, priorities had shifted twice, and three half-started ideas from the month before sat abandoned in a folder nobody opened. Lena thought of herself as a source of inspiration. Her team saw her as a source of chaos.

She didn't have bad intentions, and she cared deeply. What she missed was the hard work underneath, aligning ideas with real problems, real people, and a pace the business could absorb.

The team smiled politely, said "Sounds great," and waited for the next idea to blow in from seat 6A.

The last chapter was about the culture that lets a team take up AI. This one is about what you do to lead it, and Lena is the warning at the front of it. She gave her team energy, tools, and enthusiasm. What she never gave them was direction.

Features Are Not Strategy

Here's the confusion at the heart of it. A lot of leaders believe that bringing in tools is the same as leading the shift to AI. The newer the tool, the more current the buzzword, the more it feels like progress. It isn't. Handing your team a tool is not the same as giving them a direction, and the two can even work against each other. Every new tool Lena carried back cost her team a week of momentum on the last one.

Underneath it is a feeling you probably know. Everyone is talking about AI, competitors are supposedly racing ahead, and every week brings a new tool that sounds essential. The fear of being left behind is real, and it pushes you to do something, anything, right now. Grabbing the latest tool feels like action. It scratches the itch. That's exactly why it's dangerous, because motion and progress feel identical in the moment and only look different months later, when you see how far the business actually moved.

Leading AI in your business starts one step back from any tool. It starts with knowing where the business is going. A tool is only useful in service of a goal, and if the goal isn't clear, no tool supplies one. This is worth saying plainly, because it's the thing most often skipped. You cannot build a successful AI strategy on top of a business that doesn't have a strategy. AI doesn't set your direction. It moves you along the direction you've already set, faster. Point it at nothing in particular and it will take you nowhere faster.

Once the direction is clear, the job isn't to pile on tools. In a business your size it's usually the reverse. What your team needs is fewer, chosen well, and the room to get good at them. Every tool you add carries a cost that never shows up on the invoice. Someone has to learn it, fit it into how they already work, and decide when to reach for it over the three other things that do something similar. Stack enough of them and your team spends more energy managing tools than doing the work. The mess Lena created wasn't that any single tool was bad. It was that nobody could hold all of them at once, so none of them ever got good.

Part of the job is restraint. Protect your team's focus. Say no to most of what crosses your desk, so the few things you say yes to get a real chance to land. The other part is backing the ones that make the cut, which is the part Lena never reached because she was already onto the next idea. Pay for the proper version instead of the free tier that quits at the first limit. Give people time to learn it, on the clock, not squeezed into their evenings. Use it yourself, in front of them, so they see it's real and not one more thing announced and forgotten. Make it safe to come to you when it doesn't work, because sometimes it won't. That backing is what turns a tool from an announcement into a habit. Without it, even the right tool dies, and your people learn that the smart move is to wait out whatever you bring next.

Transfer or Transformation

Most AI in a small business moves work around without changing much. We see it constantly. The work gets shuffled from a person to a tool far more often than it gets genuinely rethought, and the difference between those two is where you come in.

Here's why it falls to you. Your team, heads down in the daily work, will almost always point a new tool at the task in front of them. That's the natural move, and it produces the shuffle. Seeing the other option, the one where the work itself changes shape, means stepping back from the daily grind to ask a bigger question, and stepping back is the part of the job that's yours. Your team needs you to be the person asking it, because they mostly won't get there on their own.

Take the automated note taker, which every business seems to be adding right now. Think about what happened before. People met, and often nobody took real notes, or minutes got written and never sent. Now a tool transcribes the whole meeting and drafts the minutes. Real capability, and it feels like progress. Here's the question a leader has to ask, and the team usually won't. Does anyone read them? If the minutes land in a folder nobody opens, exactly as before, nothing actually changed. You've automated the production of a document nobody wanted. The meeting is still the meeting, the follow-up still doesn't happen, and the tool works perfectly while the outcome stays identical.

That's a transfer. The task gets done by a machine instead of a person, and the underlying result sits exactly where it was. It's the most common thing that happens when a small business adopts AI, and it's easy to wave through as a win because there's a working tool to point at.

Transformation is a different order of thing, and it rarely happens by accident. It's when the way the work gets done actually changes. Your clients start booking, quoting, and answering their own questions through a system that handles it end to end, and a whole category of calls stops coming in. A job that used to need three handoffs now needs none. Something that took a week happens while the customer waits. That kind of change almost never comes from the person doing the task, because they're focused on doing it well. It comes from someone with the distance to ask why the task exists at all, which is you.

Understand Thy Problem and Communicate It Well

Transformation is not something you buy. It's a change in how the whole organization thinks about its problems, and a commitment to learning your way through them together. The tools are the easy part. The hard, valuable part is building the habit of looking at a real problem, understanding it deeply, and getting everyone aligned on what would actually fix it. That habit is what separates a business that gets stronger with AI from one that just accumulates subscriptions.

It takes commitment, and it takes education across the whole team, not just the person at the top. When people understand how to frame a problem and think it through, they stop waiting to be handed a tool and start bringing you real opportunities. Here's a simple framework that helps you build that habit together, five questions that turn a problem worth solving into a project you can run.

Answer them in order, as a team. By the time you reach the fourth question, you already share a clear picture of the problem, what success looks like, and who's involved, so the work you plan is aimed at something real. Use it whenever a problem is worth solving, and bring it into your quarterly and yearly planning, so thinking this way becomes part of how the organization sets direction.

Why does it matter? Start with the problem, tied to a real number, and be honest about what's causing it. Say a dental practice works out that a chunk of its lost revenue comes from empty chairs, patients who don't show and slots that can't be filled in time. That's the reason. A number that hurts, and the reason behind it. If you can't put a figure on it, it isn't ready to build on yet.

What would success look like? This has two parts. The objective is the plain statement of the win, a schedule that stays full even when life gets in the way. The key results are how you'll measure it, no-shows down from eighteen percent to under eight, and a real dollar figure recovered each month. The objective inspires, the key results tell you whether you got there.

Who will help you get there? Name the people, and bring them in before you plan the work, because the people you have shaped what's realistic and because a change they helped design is one they'll carry. Who owns it, who does each piece, whose daily work changes, and who needs to understand it early rather than have it handed to them finished.

How will you reach it? This is the plan to make it real, and it's more than the tool. What data you need and how you'll clean it. Where AI does the work only AI can do, understanding a patient's reply, personalizing the outreach, predicting who to call first. What a human still checks before anything goes out. How you'll train the team, and how you'll pilot it small before betting the whole practice on it.

When will it happen? Put dates on it. What gets done this quarter, what comes next, and when you'll check whether the early signals are moving. A project without a timeline is a wish, and a wish always loses to the daily work. Milestones are also how you catch a stall early. If the first checkpoint slips with no good reason, you've learned something before you've spent the whole budget.

Here is the whole canvas filled in for that dental practice, the kind of problem that hides in plain sight because no single missed visit feels like much.

A Completed Canvas: Cutting No-Shows at a Dental Practice

Why does it matter?	No-shows run at about 18% of appointments. Each empty chair is roughly $200 that can't be rebooked on short notice, close to $8,000 a month. Reminder calls reach voicemail, and nothing fills a slot that opens up the day before.
What success looks like?	Objective: a schedule that stays full even when life gets in the way. Key results: no-shows from 18% to under 8%, half of same-day cancellations filled from a waitlist, about $5,000 a month recovered over one quarter.
Who will help?	Dr. Alvarez approves the patient messaging. Maria, the office manager, owns it. The front desk, Jen and Priya, runs it day to day. Bring them in from the start, since it changes their morning most.
How will you reach it?	1. Get the patient history clean, past visits, no-show patterns, contact preferences. 2. Set up AI to write each reminder personally and read plain-language replies, so a patient can text "can't make Tuesday" and be offered a real alternative. 3. Have the AI rank the cancellation waitlist by who is most likely to take a given slot, and offer it to them first. 4. Write the guardrails: what the AI sends on its own, and what a human approves, especially anything clinical. 5. Train the front desk, then pilot on one schedule for two weeks before rolling out.

When will it happen?	Reminders live by the end of month one, waitlist and auto-offer by month two. Check the confirm rate at 30 days, the no-show rate and recovered production at 90.

Worksheet

The Project Canvas

A five-question canvas that turns a business problem into a runnable AI project.

Scan or visit **designated.ai/books/practical-strategy**

Work through those five as a team and look at what you have. A shared understanding of the problem, a clear picture of success, the right people, a real plan with the tools in their proper place, and a schedule to make it happen. Do this a few times and it stops feeling like a framework and starts feeling like how your organization thinks.

Four Gates Before You Automate

The danger with a fast, cheap tool is that you can automate almost anything now, including things that shouldn't exist. Point AI at a broken process and you get the broken process running faster, producing wrong answers at higher volume with more confidence behind them. A weekly report nobody reads shouldn't be summarized faster. It should stop being made. Before any tool touches a task, the task has to earn it. Figure 3.2 shows the four gates it passes through in order: Eliminate, Simplify, Automate, Delegate.

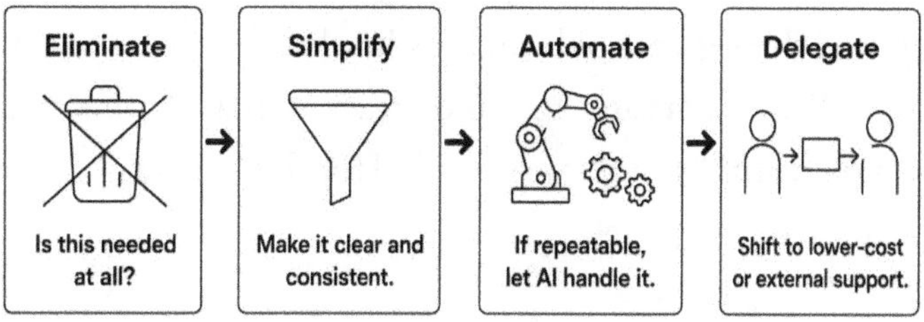

Figure 3.2. The ESAD Filter, the four gates every task passes through before a tool touches it.

Eliminate. Does this task need to exist at all? Who uses what it produces, and what decision changes because of it? If nobody can name a person who needs it or a decision it drives, stop here. You've just saved more than any tool would have, because the fastest version of a task is the one you no longer do.

Simplify. For what survives, strip it down. Which steps are there because of a problem that no longer exists, a workaround for an old system, a check nobody remembers the reason for? Cut those first. A task carries years of barnacles, and you don't want to automate the barnacles.

Automate. Only now do you ask whether a tool should take it. The real question is whether the task is consistent enough that the same instruction works every time. If every case is a judgment call or a special exception, it isn't ready, no matter how good the demo looked. Consistency and repeatability can be handed over.

Delegate. If it isn't a fit for a tool, ask the last question before you build anything. Could a person do the simplified version faster or cheaper than you could build and maintain an automation? Sometimes the honest answer is yes, and a five-minute human task isn't worth a week of setup. Delegating to a person, or to a capable tool that genuinely handles it, both count. The point is that building something isn't always the answer.

Run it on paper, one task per page, and you'll notice something. Most tasks never make it to the automate gate. They get eliminated or simplified out of existence first, which is the whole point. The instinct is to jump straight to automating, and these gates hold you back just long enough to ask the cheaper questions first. Your team can run this on their own work once you've shown them how, and it turns "should we get a tool for this" into a conversation with a clear order to it.

Worksheet

The Four-Gate Worksheet

A short playbook to evaluate before automate.

Scan or visit **designated.ai/books/practical-strategy**

Design the Role Before You Fill It

The clearest place to practice this is the moment you're about to hire.

The reflex, when the work piles up, is to hire someone to do it the way it's always been done. You write the job description from the last person who held it, or from the pile of tasks that's overflowing, and you look for someone to take it off your plate. That's the same reflex as pointing a tool at an existing task. It fills the gap without asking whether the gap should exist in the shape it's in.

Before you write that job description, run the transformation question on the role itself. If you were designing this position from scratch today, knowing what these tools can do, what would actually be in it? Some of what used to fill a person's week, the sorting, the data entry, the first draft of routine documents, the chasing of information, may not need a person at all anymore. What's left is often the part that needed a human all along, the judgment, the relationships, the decisions, the handling of the exceptions a tool can't.

That reframing changes who you hire and what you pay for. A role redesigned around AI is usually a better job, more judgment and less grind, and it often needs a different kind of person than the old version did. You might find you need fewer hands doing routine work and one sharper person owning the outcome. You might find the role you were about to post doesn't need to be a full hire at all, and that a current employee, freed from the grind by the same tools, could grow into the interesting half of it.

This is where a leader earns the most leverage, because it's a decision only you can make. The person doing the job can't redesign their own role, and a job description copied from last year quietly locks in the old way for another few years. Every hire is a chance to build the work around what's actually possible now, or to cement the way things used to be done. You're the only one standing at that fork.

None of this means treating people as line items to cut. The point isn't fewer people for its own sake. It's that when you design a role around what humans are genuinely best at and let the tools carry the rest, you get better jobs, better work, and people who spend their days on things that matter. That's the same promise from the last chapter, that improving the work makes people more valuable, seen from the other side of the hiring desk.

Keep Learning, Out Loud

Everything in this chapter asks you to see a little further than the person doing the daily work, to spot the transfer, to picture the redesigned role, to ask the bigger question. You can only do that if you keep learning what these tools can actually do, because the ground moves and last year's sense of what's possible goes stale fast.

Your team takes its cue on this from you. If the person at the top stopped learning the day the business found its footing, that becomes the ceiling for everyone. If you're visibly still curious, still trying things, still willing to say a tool surprised you or embarrassed you, that becomes the norm instead. Learning out loud is contagious in a small business, and so is its absence.

You don't need to become technical. You need to stay current enough to lead the conversation, to know roughly what's possible so you can tell a real opportunity from a shiny distraction. Try the tools yourself on your own work. Share what you find with the team, the wins and the flops both. Make a place, even just a shared folder, where people can drop what worked for them so a good find helps everyone instead of one person. The habit matters more than the format.

If you've read this far into a book like this one, that habit is almost certainly already yours. You're the kind of leader who invests time in getting this right. The rest is practice.

What Teams Actually Want

Lena is more common than she sounds. Most leaders who lose their team this way meant well, bringing back the latest thing they'd heard, sure they were helping. What teams actually need is quieter than that. Clarity, steady support, and a leader willing to say, "We're figuring this out together."

In a small business, change is personal. A new tool doesn't just touch workflows. It shifts trust, control, and the routines people have relied on for years. Watch what people ask once the demo ends and the leader leaves the room. What's expected of me now? What's changing and what's staying the same? How do I keep up? Am I still valuable here? Those questions decide whether AI takes hold, and not one of them is about the tool.

A team with clear answers will try almost anything. A team left guessing will smile through the demo and quietly go back to the old way, because the old way is the only one that was defined. Nobody expects you to have every answer. What they need is a leader who has set a direction, who aims for real change instead of motion, who designs the work around what people are best at, and who keeps learning alongside them. Do that, and the four questions answer themselves. That's what your team needs from you, and it's what turns a tool everyone forgets into a change that lasts.

All of that raises a practical question you'll face the moment you're ready to act. Of everything AI could touch in your business, what do you actually start with? The first thing you choose matters more than it looks, because it teaches your team what AI is going to mean for them. The next chapter is about picking that first move well.

Chapter 3.3
Pick the Right Use Case

The first time a team watches AI do something genuinely helpful, there's a spark. A long email thread becomes three clean bullet points. A report that normally takes half a day appears in minutes. A messy list of meeting notes turns into clear actions. The room feels slightly lighter, as if someone quietly opened a window. And then the ideas start rushing in:

With a clear understanding of what your team needs to succeed with AI, the next step is to decide where to focus those strengths. In the following section, you'll learn how to pick the right first use case. It should deliver real value, build trust, and set the stage for future progress.

"Could it do this for the whole department?"

"What about onboarding?"

"Could we automate our customer support next?"

"Let's rebuild our entire process around AI."

This moment is exciting and dangerous. Early AI initiatives can become difficult to implement when their scope expands before requirements, stakeholder expectations, risks, and oversight mechanisms have been clearly defined. [6] The first use case becomes a giant project with fuzzy boundaries. Teams debate data access, security, ownership, and edge cases. Weeks pass. The pilot goes quiet.

Meanwhile, the rest of the organization watches and waits. When early AI initiatives fail to demonstrate clear value or create excessive operational complexity, they can make it harder to build user confidence in subsequent deployments.

The technology may still be powerful, but belief in it is beginning to wane.

The first visible AI use case can therefore have an important influence on how employees perceive subsequent AI initiatives. It tells people whether AI will relieve pressure or add another complex system to learn. It sets the speed, the tone, and the confidence level for everything that follows.

So the question isn't: "What's the biggest thing AI could automate?"

It's rather: "What's the smartest first thing to automate, so people feel the value fast and want the next step?"

For a practical first deployment, this means selecting a bounded task or small workflow in which AI can provide observable value while risks, outputs, and human responsibilities remain manageable.

The 5-Hours Effort Saving Rule (5HESAR)

In this book, we propose the 5-Hours Effort Saving Rule (5HESAR) as a practical heuristic. Select an initial AI use case that has the potential to save approximately five hours of human effort per week while remaining bounded, reviewable, and low risk.

The reasoning behind this rule is practical. Five hours is large enough for employees to notice during an ordinary workweek, while still encouraging teams to begin with a bounded problem rather than a broad transformation program.

A strong first use case should not be seen as the end goal. It's the first brick in a much larger foundation. It should therefore be small enough to implement successfully and deliver results quickly, yet useful enough to create immediate and visible value. It must also be safe and manageable, allowing problems to be corrected without serious consequences. Finally, it should be simple and accessible so that the barriers to adoption remain low. By meeting these criteria, the first use case can build confidence, generate momentum, and create a solid basis for future development.

In summary:

You're not choosing the most advanced use case. You're choosing the most adoption-ready use case. Early on, the real deliverable isn't just automation, it's trust.

Think of AI adoption like moving into a house with smart home features. You could start by rewiring everything at once. Lights, heating, security, appliances, and voice control. But if the first week is full of glitches, resets, and confusing controls, the household quickly decides the old way was better.

Most people don't start that way. They begin with a simple upgrade that improves daily life immediately, such as smart lights in the living room or a thermostat that learns routines. It's easy to test, easy to reverse, and the benefit is obvious. After that first win, everyone becomes open to the next upgrade.

Your first AI use case works the same way. It should feel like a small improvement that makes everyday work smoother, not a radical remodel that throws people into "construction mode."

Start with one room that matters. Let people experience the comfort and then expand.

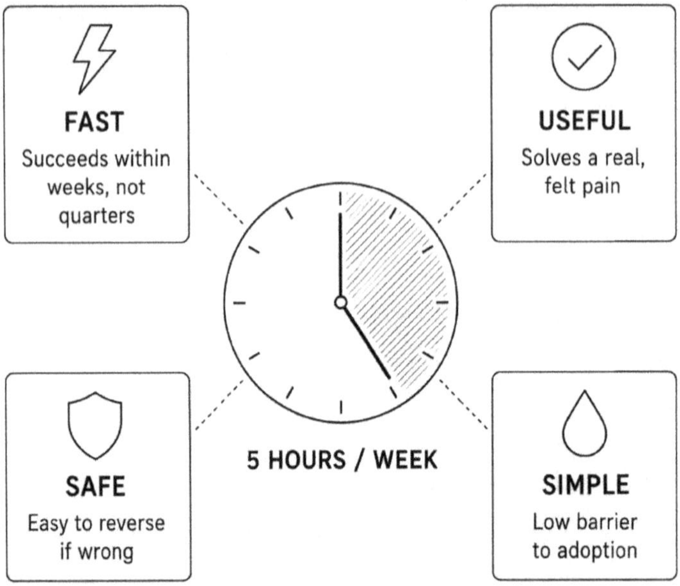

Figure 3.3. The 5 Hours Effort Saving Rule (5HESAR)

Worksheet

The 5HESAR Picker

Score any candidate's first use case in five minutes.

Scan or visit **designated.ai/books/practical-strategy**

Industry Applications

Here are practical first-use-case examples by role, designed to fit the 5HESAR, see Figure 3.3. and serve as building blocks for longer-term workflows.

Owner / Executive Roles

Executives need AI to clear the fog faster, not to replace judgment but to reduce noise. Their pain usually sits in information overload and decision preparation.

Strong early wins would be:

- Weekly business summary agent pulls key metrics, highlights anomalies, and drafts one-page leadership updates.
- Meeting prep + follow-up assistant summarizes prior discussions, drafts agendas, and extracts action items afterward.
- Decision log builder captures decisions and rationales to prevent context loss across weeks.

They fit the rule because they provide high pain (too much info), low risk (reviewable), and a fast payoff (immediate clarity).

Operations Roles

Ops teams are buried in coordination. Their pain isn't due to skill, it's time lost to chasing, syncing, and updating.

Strong early wins are:

- Auto-generated status updates, AI reads project tools + communications and drafts progress summaries.
- Workflow nudges, reminders for missing inputs, deadlines, or blocked handoffs.
- Internal request triage sorts routine tasks, tags urgency, and routes to owners.

They fit the rule because they address a high-friction coordination burden, remain safely reviewable, and deliver rapid day-to-day time savings.

Marketing Roles

Marketing pain hides in execution volume: drafting, reformatting, repurposing, and reporting are all necessary, all repetitive.

Strong early wins:

- First-draft content generator with blogs, emails, social posts, ad variants that humans refine.
- Repurposing workflow which automatically turns one long asset into multiple short formats.
- Campaign reporting assistant which summarizes performance data and drafts learnings.

Why they fit the rule, because they are high pain (throughput load), low risk (humans approve), fast payoff (output accelerates fast).

Finance Roles

Finance pain is repetitive and deadline-driven. Month-end never waits, and manual errors are exhausting.

Strong early wins:

- Expense categorization and anomaly checks with flag duplicates, policy mismatches, and unusual vendors.
- Close-prep assistant to gather invoices, check completeness, and draft reconciliation packets.

Why they fit the rule, because they are high pain (manual grind), low risk (review standard), fast payoff (hours saved each close).

Traditional automation has been highly effective for well-defined processes with predictable rules and structured inputs. Many knowledge-work activities, however, involve natural language, documents, messages, and other less structured information.

What has changed is that modern AI systems can process natural language, extract information from unstructured text, generate summaries and responses, and support tasks that were difficult to express as fixed rules. Their outputs, however, remain probabilistic and require appropriate validation and oversight.

That unlocks first use cases that classical automation couldn't touch, the "human glue tasks" that quietly eat hours, which are summarizing, reformatting, reconciling, chasing context, and stitching systems together.

So the shift looks like this:

- Before: hours spent copying data, formatting reports, chasing approvals, and summarizing conversations.
- After: AI can automate or assist with substantial portions of routine information-processing work, while human review remains important where outputs affect consequential decisions.

The real impact isn't just time saved. It's mental bandwidth returned.

People stop spending sharp energy on dull tasks. When employees experience a useful improvement directly, it can create a stronger basis for broader acceptance than abstract demonstrations of technical capability alone. That benefit comes with a caveat. Excessive dependence on AI may also weaken critical engagement, which is why humans should remain actively involved where judgment matters.That's why the first use case matters now! AI is already capable of supporting many everyday knowledge-work tasks, but organizations need to determine where those capabilities can be applied reliably and responsibly.

The first AI use case is not only a technical pilot, it can also function as a trust pilot. In practice, early deployments give users direct experience with the system's usefulness, transparency, controllability, and limitations, all of which can influence trust in subsequent AI applications. Leaders often assume the hardest part is selecting the most powerful application. But early on, what matters more is selecting the most believable one, the one that removes real pain safely and quickly.

Pick the right first use case, and you create a flywheel!

A successful early deployment can build confidence, generate additional ideas, and create momentum for broader adoption.

Pick the wrong one, and you create drag:

Conversely, a prolonged pilot with unclear value can increase skepticism and make subsequent adoption more difficult.

Based on these considerations, our practical design rules are:

- High pain to guarantee relevance
- Low risk to protect confidence
- Fast payoff (5 hours/week) to build momentum

- Use cases by role to drive adoption where friction is real
- Start small, think long-term so every win becomes a building block

Because the smartest way to go big with AI is to go right first.

Once you've chosen a first use case that delivers a real, adoption-ready win, the next challenge is scale. The early success creates momentum, but it also raises a practical question. How do we expand this without getting stuck in complexity, cost, or tool chaos?

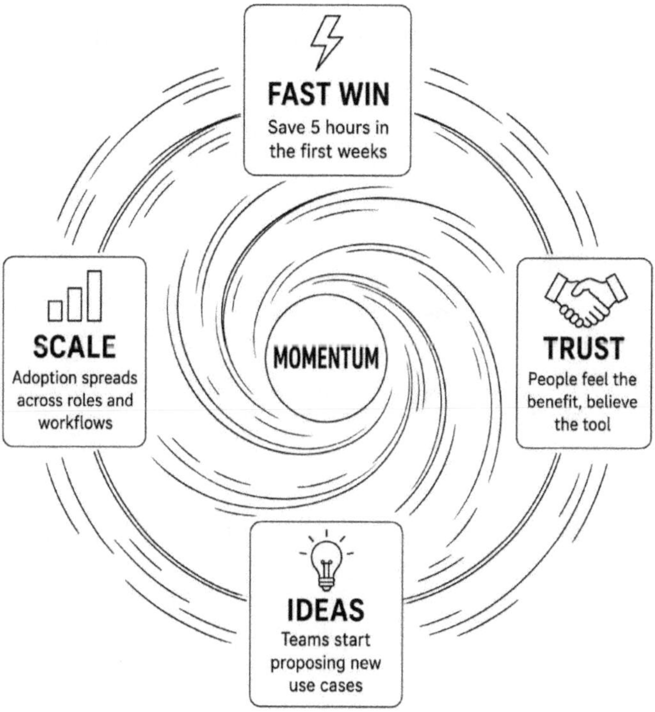

Because not every workflow should be automated the same way. Some use cases may justify tailor-made solutions, others may be addressed by established products, and some may be suitable for lightweight AI agents operating under defined controls. The key is learning to pick the right path for each new use case, so you keep speed without sacrificing control.

Promptpack

First-Use-Case Ideas by Role

Ready first moves for Owner, Ops, Marketing, and Finance.

Scan or visit **designated.ai/books/practical-strategy**

That brings us to the next decision framework: Build vs. Buy vs. Delegate.

Chapter 3.4

Build vs. Buy vs. Delegate

In the previous chapters, you defined your readiness, identified your friction points, and built a roadmap. Now you face a critical operational choice. You know what you want to solve. The remaining question is how you will solve it.

AI strategy is not only about what you want to do. It is about what you can sustain.

Most small and midsize businesses struggle not because they choose the wrong tool, but because they choose the wrong delivery path for their reality. They build custom solutions when a proven product would have been sufficient. Or they delegate a core process without a plan for ownership once the work is complete.

Before committing to a platform, a pilot, or a hire, you must decide whether you will Build, Buy, or Delegate.

This decision cannot be made on impulse. It must come after your audit. Until you understand your workflows, the condition of your data, your available capacity, and who will own the solution after launch, you do not yet have the inputs required to choose wisely.

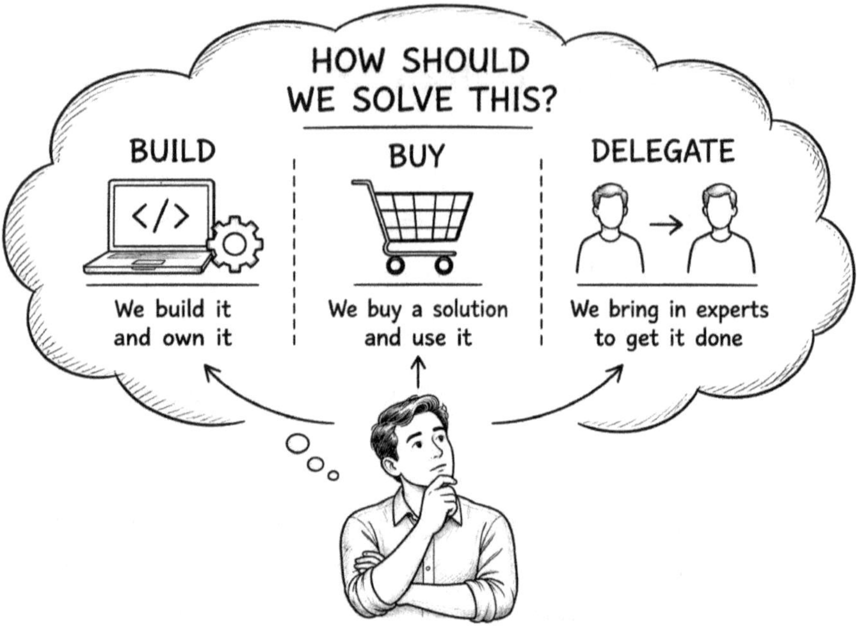

Clear Definitions

To make the right decision, we first need to remove marketing language and define the paths clearly.

Build means you develop and own the system internally. This includes custom workflows, proprietary automations, API integrations, or locally managed AI solutions. You own the code, the data, the risk, and the ongoing maintenance.

Buy means you use commercial AI tools or platforms. This includes SaaS subscriptions, built-in AI features inside your existing software stack, and packaged automation tools. You rent the capability while the vendor manages infrastructure, updates, and reliability.

Delegate means you outsource execution to a partner, freelancer, or agency. You define the outcome and remain accountable for results, while the external party handles technical implementation.

Each path has a role. The goal is not ambition. The goal is fit.

When Buying Makes Sense

For most organizations, especially in early sprints, buying is the most responsible starting point.

Buying makes sense when speed to value matters more than customization. If the problem you are solving is common: summarizing meetings, drafting content, organizing requests, routing information, the solution should be common as well.

For the vast majority of SMBs, the fastest path to meaningful progress is not building an internal system. It is using trusted, enterprise-grade tools and the AI capabilities already embedded in platforms they operate every day.

Buying is not a shortcut. It is a form of discipline. It prevents experimentation from turning into premature engineering. When teams are small, stretched thin, or still strengthening foundations, buying allows learning to happen alongside real operational impact. It reinforces a core principle from earlier chapters: tools should follow problems, not the other way around.

The primary risk of buying is not performance. It is sprawl and governance. Without clear standards, organizations accumulate overlapping subscriptions and expose sensitive data through poorly chosen tiers. This is why enterprise licenses, access controls, and usage guidelines matter.

Buying allows teams to move quickly while preserving focus and control.

When Building Makes Sense

Building is seductive. It feels like innovation. With modern AI tools, it is also easier and faster than ever to do. That is precisely why discipline matters more than ever.

Building makes sense when AI becomes part of how you differentiate, not just how you automate. If the workflow is tied directly to your competitive advantage, reflects unique data or logic, and will be reused over the long term, then building can be a smart investment.

This distinction is increasingly visible among AI-native companies. McKinsey found that organizations leading in AI tend to build when a capability creates a defensible advantage through proprietary data, expertise, or intellectual property that an off-the-shelf solution cannot replicate. For more standard capabilities, buying remains the more practical path. [7]

This is where internal systems, multi-step workflows, and deeper integrations become proprietary assets. You build when no off-the-shelf solution can reasonably support the nuances of your operations or customer experience. At this stage, you are no longer experimenting. You are deliberately creating capability.

AI dramatically reduces the cost and speed of building. It does not remove the responsibility of ownership. **The ability to build something in a weekend should not be confused with the ability to operate it for years.** Code can be generated quickly, but systems still require maintenance, monitoring, documentation, and governance. Strategy determines whether what you build survives beyond the first demo.

The real risk is rarely technical complexity. The risk is unclear ownership. When no one is responsible for updates, fixes, or evolution, even well-designed systems quietly degrade.

A custom solution without durable ownership does not become a strength. It becomes a fragile dependency.

Build deliberately. Only build when ownership, maintenance, and long-term support are clearly established.

When Delegating Makes Sense

Delegation is the bridge between ambition and capacity.

It makes sense when you want outcomes without becoming responsible for building the infrastructure yourself. This is often the right choice when AI supports operations but is not core to differentiation, or when specialized skills are needed temporarily.

Delegation works well when you need expert execution such as complex automation, system integration, or structured AI setup, but intend to use the result daily. You delegate execution, not accountability.

The risk appears when delegation is treated as set and forget. Without a clear handover plan, delegated systems fail in the same way poorly planned internal builds do. Knowledge remains external, and when the partner disappears, the workflow stalls.

Delegation succeeds only when internal ownership remains intact.

Delegate to accelerate progress, not to outsource responsibility.

Build, Buy, or Delegate: A Sustainability Lens

The decision ultimately comes down to where complexity will live.

When you build, your organization absorbs that complexity. You gain greater control and the opportunity to create something proprietary, but you also accept responsibility for maintenance, reliability, security, and evolution.

When you buy, much of that complexity shifts to the vendor. You gain speed and reduce the burden on your internal team, but you accept less control and greater dependence on another company's platform.

When you delegate, you transfer much of the execution to someone with the expertise or capacity you currently lack. This can accelerate progress, but it does not eliminate the need for internal ownership. Someone inside your organization still needs to understand the outcome, manage the relationship, and know what happens when the engagement ends.

In practice, Build, Buy, and Delegate are not mutually exclusive. You may buy an AI platform, delegate the initial integration to a specialist, and build a proprietary workflow on top of it. The decision should be made at the workflow level, not for the organization as a whole. What matters is knowing what you need to own, what you can rent, and where outside expertise can accelerate execution.

The right choice is not necessarily the most advanced one. It is the one your organization can support after the excitement of implementation is over.

The Decision Filter

When a new problem appears on your roadmap, apply this filter before committing resources:

1. Do you need ownership, or do you primarily need the outcome?

Ownership may point toward Build. If the outcome matters more than owning the underlying system, Buy or Delegate may be the better path.

2. Is this workflow strategic or standard?

Differentiation justifies Building. Common workflows justify Buying.

3. Can your team maintain it without constant intervention?

If not, Buy or Delegate with a service plan.

An AI initiative is not complete when it goes live. It is complete when it operates reliably within your daily work without requiring constant, emergency intervention.

Before you commit, evaluate your path:

- **Build** only when the value is high, the solution is unique, and you are prepared to own the long-term maintenance.
- **Buy** to move quickly and solve standard problems where the vendor handles the complexity.
- **Delegate** to bridge capability gaps without needing long-term hiring.

The path you choose matters less than your ability to support it. If you need full ownership and differentiation, building is justified. If the workflow is common, buying is usually the superior choice. If the skill gap is temporary, delegation will accelerate your progress.

Worksheet

The Build-Buy-Delegate Filter

Three questions to place any new project on the right path.

Scan or visit **designated.ai/books/practical-strategy**

Chapter 3.5

Choosing Allies, not Just Vendors

By the time you arrive at this chapter, you are no longer at the beginning of your AI journey. In chapter 3.3, you identified a concrete, valuable first use case. In chapter 3.4 you decided whether to build, buy, or delegate. Now a different question comes into the spotlight. Who do you trust to help you do this well?

For many established organizations, AI adoption does not begin with a clean slate. It is introduced into an existing technology landscape of systems, data, workflows, and organizational processes. [6,8]

Such companies already have a Customer Relationship Management system (CRM), an Enterprise Resource Planning system (ERP) and already depend on accounting, project management, as well as office tools.

The real decision most leaders face is not which AI vendor should they use, it is "Can the vendors we already trust take us to the AI era, or do we need to look elsewhere?".

This is where this chapter begins.

In this chapter, we use the term "AI ally" for a provider whose contribution extends beyond software delivery to long-term support for safe, effective, and sustainable AI adoption.

Illustrative Example

The following case study is fictional and is intended solely to illustrate the concepts discussed in this chapter. While it refers to a real organization, the scenario, conversations, and decisions described are entirely hypothetical.

Imagine Deutsche Bahn, one of Europe's largest railway operators. The company has invested for years in enterprise systems for operations, maintenance, procurement, and asset management. Engineers responsible for rolling stock maintenance work with extensive technical documentation, inspection reports, and maintenance records to keep thousands of trains operating safely and reliably. After identifying predictive maintenance as its first AI use case, the leadership team naturally turns to its existing software vendors. One vendor proudly demonstrates a conversational AI assistant capable of answering maintenance questions. Another presents sophisticated dashboards with impressive failure predictions, while a third promises a fully autonomous maintenance platform within months. Every demonstration is polished and convincing, but the leadership team soon realizes that impressive presentations alone do not answer the most important question.

Which provider will actually help Deutsche Bahn succeed over the coming years?

The evaluation focuses on the company's real operational needs rather than feature lists alone. The engineering team asks each vendor to demonstrate the solution using representative maintenance data instead of carefully prepared examples. Maintenance managers want to understand how uncertain predictions are communicated to technicians, how recommendations can be reviewed or overridden, and how human approval is incorporated before maintenance schedules are changed. The IT department examines how well the solution integrates with existing SAP systems, how data can be governed and exchanged securely, and whether the proposed architecture fits the company's long-term technology landscape. At the same time, training managers ask what happens after deployment. Will engineers receive practical training? Will user feedback influence future improvements? Will experts remain available when the first unexpected situations arise?

As the evaluation progresses, the differences become increasingly apparent. One vendor consistently explains not only what the AI system can do, but also where its limitations lie. Rather than promising perfect predictions, the team openly discusses model uncertainty, demonstrates human review mechanisms, explains governance features, and recommends starting with a carefully controlled pilot before expanding. Another vendor continues to showcase impressive demonstrations but struggles to answer questions about governance, error handling integration, and long-term operational support. The distinction isn't just about technology anymore. One company is primarily selling software, while the other is preparing to become a long-term partner in Deutsche Bahn's AI transformation.

Deutsche Bahn ultimately selects the second provider as its AI ally. The distinction is not based solely on model performance or licensing costs, but on the provider's willingness to understand operational workflows, acknowledge limitations, support responsible governance, transfer knowledge to internal teams, and continuously improve the solution together with its users.

Over time, engineers begin trusting the system because they understand both its strengths and its limitations. The first AI application succeeds, not because it delivers perfect predictions, but because it is introduced responsibly, integrated into existing processes, and supported by a provider committed to long-term collaboration rather than a one-time software sale. That is the difference between purchasing AI software and choosing an AI ally.

Although this introductory example comes from large enterprises, the same principles apply to small and medium-sized businesses. The difference is not the decision itself but the scale. An SME may not operate thousands of trains or maintain a large IT department, yet it faces many of the same questions. Should we rely on our existing ERP or CRM provider? Can this solution integrate with our current system? Will our employees receive training? Who supports us once the software goes live? And will this provider help us expand AI to additional business processes over time?

For smaller organizations, these questions are often even more important because they typically have fewer technical resources and less room for costly mistakes. Choosing the right AI ally therefore becomes less about selecting the most advanced technology and more about selecting the partner who will help the organization build confidence, capability, and value.

With this illustrative example in mind, we will now work through the key considerations for choosing the right AI ally. The goal is not simply to select a vendor, but to find a partner who can support your organization throughout its AI journey.

Hence, treating AI tools and providers as long-term strategic partners rather than one-off purchases, regardless of whether they are long-standing incumbents or new market entrants is very important. Traditional software procurement typically focused on functional requirements, pricing and licensing, implementation effort, and ongoing maintenance.

AI changes that. AI-enabled systems may exhibit probabilistic behavior and may be updated over time as models, data, or system configurations evolve.[9]

The right AI vendor helps you reduce risk, co-design use cases, support adoption, and leave your organization more capable than it was before.[6,8] That makes them an ally. This applies equally to existing vendors and new vendors. Longevity alone does not make an ally. Capability, honesty, and commitment do.

Think of AI adoption like a long hike. A vendor is someone who sells you fancy boots at the trailhead. They look great. They promise performance, but once you're on the mountain, you're on your own. If your feet hurt, the weather changes, or the terrain turns unfamiliar, you quickly find out whether those boots were made for your journey.

An ally is different. An ally hikes with you. They know the terrain, help you adjust your pace, remind you to hydrate before you collapse, and if you twist your ankle, they don't disappear after the sale. They help you reach the summit safely.

AI adoption is a long hike, not a one-time purchase. The question is not who has the coolest boots, but who will help you finish the trail. With existing vendors, you often start with more trust and familiarity. But even trusted partners can fall behind as the terrain changes. Familiarity is helpful until it turns into inertia.

Evaluating Existing Vendors Through an AI Lens

For most SMBs, the first step is not to search for new AI platforms. It is to evaluate existing vendors through an AI lens. Key questions to ask include how fast they respond to the generative AI wave, or whether AI features are deeply integrated and not feel bolted on, or do they solve your actual use case without just demonstrating generic capabilities. Is there a credible roadmap, rather than just marketing language, or whether you can access and export your data cleanly if you need to switch later. The final key question would be whether they support governance, human oversight, accountability, provide adequate training, documentation, and knowledge transfer so that internal users can understand and govern the system effectively. [6,8,10]

Loyalty is comfortable, but loyalty without evaluation becomes a liability when the market is moving this fast. If an existing vendor passes this test, staying can be a smart and efficient choice. If they fail, they should be treated as one option among many, not the default.

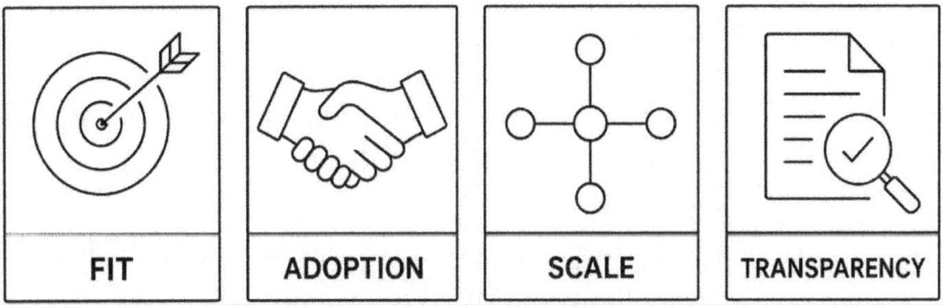

Figure 3.4. Evaluation criteria for existing vendors or new ones

Evaluation criteria

In Figure 3.4, whether you are evaluating existing vendors or new ones, the criteria remain the same:

Fit the real use case, not a fantasy future

- ☐ Ask vendors to walk through your workflow, not theirs. Where possible, evaluate the solution using representative data and workflow conditions rather than relying only on vendor demonstrations. [6] A strong signal of an ally is excitement and openness about solving your pain instead of showcasing polished demo scenarios.

Support adoption, rather than just installation

- AI systems are unlikely to deliver sustained organizational value if users cannot incorporate them effectively into real workflows. [8] Vendors should be able to answer how the rollout will be tailored to our team, what about the onboarding of non-technical users. Moreover, what changes are required from management, how are feedback loops established, how is human oversight handled, and what happens the day after going-live?

- Here, look for evidence of sustained adoption, not just model accuracy.

- Also of growing importance is whether they provide Model Context Protocol (MCP) connectors to allow for accessing data sources across the company.

Grow across roles and workflows

- Prefer solutions that can be integrated into the existing technology environment while supporting appropriate governance and lifecycle management.[6] Consider also whether future expansion would require disproportionate redesign of established workflows. Vendors should articulate how success in one team expands to others.

Be honest about limits and risks

- ☐ Avoid vendors who pretend AI is perfect. Ask about known failure cases, hallucinations, output validation, and data-protection risks.[10,11] Acknowledging limitations and explaining how they are managed can provide a stronger basis for informed trust than presenting AI as error-free. Only vendors, existing or new, who meet these criteria should be considered allies.

Checklist

The Vendor Evaluation Checklist

Eight questions every AI vendor should answer clearly.

Scan or visit **designated.ai/books/practical-strategy**

Stay-or-Leap Trade-Off

One of the hardest decisions in AI adoption is whether to stay with existing vendors or leap to new ones. From a management perspective, staying with an incumbent may preserve existing integrations, knowledge, and investments, while moving to a new provider may offer access to different technical capabilities or architectures. Neither option is inherently superior. As a practical procurement consideration, organizations should also account for existing investments and switching costs when comparing incumbent and new providers.

A hybrid architecture may also be considered, in which specialized AI capabilities complement rather than replace established core systems.

If ROI is already achieved, your freedom increases. Staying should then be an active choice, based on trust and capability, not habit. Not every leap is smart, but not every stay is safe.

AI-enabled systems may change over time as models, data, configurations, integrations, and degrees of autonomy evolve. Models evolve, autonomy increases, and tools that feel like assistants today may become core workflow engines tomorrow. Provider selection can therefore influence not only implementation but also long-term governance, integration choices, internal capability, and future system evolution.

At the same time, the market is noisy. Strong benchmark or demonstration performance does not by itself establish effectiveness in a real organizational workflow. Long-term reliability, governance, integration, and adoption must also be evaluated in operational settings. Some tools shine in showcases but break in real workflows, and others cannot scale beyond pilots.

That is why choosing allies matters more than ever!

If you ask the right questions, you'll be able to identify your ally. It can be an existing vendor, a new vendor or a mixture from both. The most suitable arrangement may involve an existing provider, a new specialist, or a combination. The decision should be based on use-case fit, governance, integration requirements, internal capability, risk, and long-term flexibility rather than vendor familiarity alone.

There are several points which we recommend you to ask for:

Topic	Question
Feature List	What features do they have?
Implementation velocity	How fast can you implement it?
Price	What are the initial costs, and what are the annual costs for maintenance?
Internal training offerings	Do you have dedicated training offerings?
Workflow understanding	What do you know about our workflow, and how do you adapt to it?
Limits and risks	Are there any known limits or risks involved once we start the process?
Scalability	Will the offered tool still fit when we start to scale?
Internal Capability and Dependability	Does the provider help us build internal capability, or does our dependency on the provider increase over time?

Scorecard

The Ally or Vendor Scorecard

Rate your current providers through an AI lens.

Scan or visit **designated.ai/books/practical-strategy**

Even the following seems to be repeated, it needs to be reiterated again:

In AI, advantage does not come from having the most tools, it comes from having the right allies. Choose vendors who make you stronger rather than stuck. And once you've chosen them, the next question arise:

How do you prove they're working?

That's exactly what the next chapter tackles. Measuring ROI, even without a CFO.

Chapter 3.6
Measuring ROI

After you have selected the right partners to support your AI journey, the next important step is to show that these choices are delivering real results. In the following section, you will learn how to measure the true return on your AI investments, so you can move beyond activity and demonstrate outcomes that matter for your business.

Luca had never felt more modern. Six months of real adoption, and the evidence was everywhere he looked. Every client presentation now came out of a frontier generative model, one of the big general-purpose AI systems his whole team could use, and the decks looked like an agency had built them. Emails tagged themselves as they arrived, and half of them had a drafted reply waiting before anyone opened the thread. The CRM highlighted which deals were warming up and which were going cold. The intake forms routed themselves. Walk through the office on any Tuesday and there was a screen somewhere showing AI doing something.

At the quarterly sit-down, Maria, the fractional CFO he shares with four other businesses, listened to the update and asked one question.

> **Maria:** "What has it been worth, in dollars?"
>
> **Luca:** "Worth?" *He sat back. Six months of work, and she opens with an audit.* "The presentations alone. You've seen the decks we send out now. The email triage. Nobody's digging through the inbox anymore. The CRM flags the deals going cold before we lose them. The meeting notes..."
>
> **Maria:** "That's what it does." Her pen hadn't moved. "I'm asking what it's worth. Which line moved? Payroll's where it was in the spring.

Revenue too. Show me the line."

Luca opened his mouth and found he had nothing to put there. Everything on his list described motion. Nothing on it was a number she could write down.

He knew the six months were real. On the drive home he understood the problem. He had mistaken movement for progress.

This is easy to do. The tools are working, the office feels busier and sharper, and none of it has shown up where Maria is looking. This chapter is about the gap between what you can feel and what you can count, and how to close it.

ROI Means Money

As the earlier chapters laid out, before we put any AI strategy to work, the business itself needs a strategy. A strategy isn't a mood or a vague sense of direction. It sets real goals, the kind you can measure. Grow revenue by X percent this year. Scale the customer base from X to Y. Cut the cost of serving each customer by X. Open a second location by a set date. Take on X percent more work without adding headcount. Each one is a number with a target on it, which means you can tell whether you hit it.

There's a useful line to draw here. Plenty of AI use is just everyday productivity, someone drafting emails faster, cleaning up a document, getting unstuck on a first draft. That's fine, and it doesn't need to justify itself against a business goal any more than a faster laptop does. The measuring conversation is about something bigger. When you set out to make AI transformational, a real implementation you're investing time and money in, that effort has to connect to a concrete goal, or you'll have no way to tell whether it worked. Luca's mistake was subtler. He treated a pile of scattered productivity gains as if they added up to a strategic return, and when Maria asked which line moved, there wasn't one.

That's what makes her question the right one. "Which line moved" is really asking whether the investment served the strategy or just added activity around it. Return on investment means money you can find in the accounts, a cost that fell or revenue that rose. The moment you let "it feels more efficient" stand in for that, you lose the one test that tells you whether the tool earned its keep, and you keep funding tools on a hunch while the ones losing you money go unnoticed.

The honest complication is that saved time does not turn into money on its own. The clearest evidence comes from a large study of what AI adoption actually did to pay and hours. Researchers tracked twenty-five thousand workers across seven thousand Danish workplaces in 2023 and 2024. The workers reported saving roughly three percent of their time with AI chatbots. Two years of payroll records later, hours and earnings had not moved, and the data was precise enough to rule out any effect larger than two percent [12]. Twenty-five thousand people saved time, and almost none of it reached the accounts, because nothing downstream was built to catch it.

The rest of this chapter gives you a simple, repeatable way to catch that value and measure what your AI work actually returns.

Early Signals and Landed Money

Two kinds of numbers will cross your desk from here on, and it's important to tell them apart, because one is money you already have and the other is only a chance of money to come.

A landed number is money already in your accounts. Overtime that fell. An invoice that went out. A refund you didn't have to issue. It's certain, and it survives an accountant's stare.

An early signal is a probability, not a certainty. It's something you can see now that raises the odds of money later. Calls answered faster. Quotes out the same day. More leads in the pipeline. Each one improves your chances, and none of them is money yet. A faster quote is not a closed sale, it's a better shot at one. Treat every early signal as a probability the business has improved, not a payment it has received, until your own results show how often it actually turns into money.

Here's where that bites, and AI makes it easy to walk right into. Say you use AI to ramp up your marketing, more posts, more outreach, more content going out than your team could ever have produced by hand. The leads climb, fifty percent in a month, and it looks like the tool is paying off.

Whether it actually is depends on a number sitting behind the lead count, how often a lead becomes a paying customer. Volume and quality don't always move together. A flood of AI-generated content often pulls in people who aren't a good fit, curious clickers rather than buyers, so the lead count rises while the share that converts falls.

If those extra leads close at half the old rate, the pipeline number was hiding a problem, and the extra volume is worse than useless. Your team now spends its hours chasing people who were never going to buy, time stolen from the leads that would have closed. If you're paying to distribute all that content, your cost per real customer just went up.

When an early signal is a volume number, the rate behind it is the first thing to check, because more of something that converts worse doesn't just fail to help, it drains the people and the money you have.

Check It Before You Trust It

Before you lean on any early signal, make sure it actually runs ahead of your money, rather than just feeling like it should.

This is a solved problem in businesses that measure carefully. A group of researchers studied a hotel chain to see whether customer satisfaction scores could predict revenue. They could. When satisfaction scores rose, revenue didn't rise with them, it rose about six months later. That gap is what made satisfaction useful, it gave the business half a year of warning that revenue was coming. They only trusted the six-month figure because they tested it against years of the chain's own records and confirmed it actually held [13]. The delay was something they measured, not something they assumed.

The same research area shows how this goes wrong. At one large company, managers built a scorecard around measures they assumed would predict results, kept leaning on the ones that predicted nothing, ignored the ones that did, and eventually threw the whole scorecard out [14]. They never checked whether their early signals actually led anywhere. Assuming a number is predictive doesn't make it so.

So check yours. Look back through your own records and ask whether this measure has ever moved before the money did, and by how long. If you can see it in your history, you have a real early signal and a rough sense of its delay. If you can't, what you have is a hypothesis, which is fine to act on as long as you remember that's what it is.

Worksheet

The Signal-to-Money Lag Calculator

A short playbook to evaluate before automate.

Scan or visit **designated.ai/books/practical-strategy**

Give It Time to Land

By now you've picked a first project the way the earlier chapter laid out, something small and bounded, close to the work, with a quick payoff. This is about knowing when to expect the money from it, so you don't call it a failure before it's had a chance to show up.

As Figure 3.5 shows, every improvement has a delay between the signal and the money. The signal moves first. The money follows. How long it takes depends on the work. A tool that speeds up your quotes only turns into revenue when those quotes close, so if your deals usually take six weeks to sign, the money is six weeks behind the faster quotes, at least. A tool that cuts the cost of serving a customer might show up in next month's numbers. A tool that improves customer satisfaction could take two quarters before it shows in repeat business.

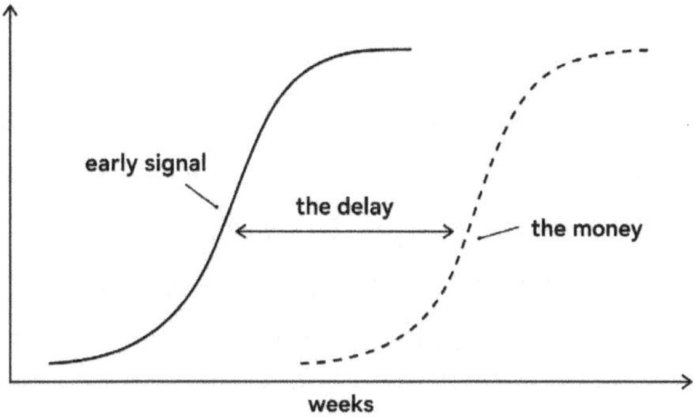

Figure 3.5 The signal moves first; the money follows after a delay.

Work out that delay before you start, from your own history. How long does a lead take to become a paying client? How long from a happier customer to a bigger reorder? You don't need precision. You need a rough window, so you know whether you're looking too early.

This is why the small, fast-payoff project you chose first matters so much. A short delay means you see the money quickly, which proves the whole approach to your team and to anyone funding it. That proof earns you the patience to run a slower project later, the kind that might take a year to pay off. Without an early win, that year is spent asking people to keep believing in faith, and faith runs out quick without real impact.

Three Levels of Proof

Every claim about what AI returned sits at one of three levels, and each one is harder to argue with than the last. Knowing which level you're on keeps you honest, and it tells you how much weight a claim can actually bear.

Noticed. Someone feels a difference. The afternoons are calmer. The team stopped dreading Mondays. The office runs smoother. This is real, and it matters, but it's the weakest kind of proof, because you can't put it in front of anyone who wasn't in the room. A skeptic nods and asks what it was worth, and you have nothing to hand them.

Counted. You have a number, measured the same way over time. Quote turnaround dropped from three days to four hours. Callbacks fell by sixty percent. The team handles thirty percent more orders a week with the same headcount. This is far stronger, because it's a fact with a method behind it, not an impression. It still isn't money, but it's the thing that turns into money.

Banked. The money shows up in your accounts and you can point to the line. This is the level Maria cares about, and it's the one that ends the argument.

Here's what banked looks like when it's big enough to matter. A contractor who cut quote turnaround from three days to same-day didn't just feel faster. Same-day quotes had always won at a higher rate, and over the next two quarters his win rate on quoted jobs climbed from about one in four to one in three. On the volume he was already quoting, that worked out to roughly a fifteen percent lift in booked revenue, without spending a dollar more on marketing. That's a number a bank would lend against.

The same win can be told at all three levels, and it gets more convincing as it climbs. At the noticed level, "quoting feels so much faster now." At the counted level, "we turn quotes around same-day instead of in three days." At the banked level, "our win rate went from one in four to one in three, a fifteen percent lift in booked work." Same improvement, three levels of proof. Your job is to push every important claim as far up as it will go, because the higher it climbs, the more it can carry.

The dip you should expect

There's a trap between counted and banked, and knowing about it in advance will save you from quitting too early. When a business adopts something new, performance often gets worse before it gets better. The old way is disrupted, people are still learning, and for a while the numbers sag. Researchers who studied manufacturers adopting AI found exactly this shape, output and profit dipping in the short run, with the businesses that pushed through coming out stronger later. Among older firms, about a third of the early loss traced to their own routines breaking down during the change, not to the technology failing [15]

As Figure 3.6 shows, economists have a name for this shape, the J-curve, because performance dips down before it swings up [16]. The dip is a normal disruption, not a verdict. If you don't expect it, week six looks like proof the whole thing failed, and you pull the plug right before the upswing. If you do expect it, you hold your nerve and let the curve finish.

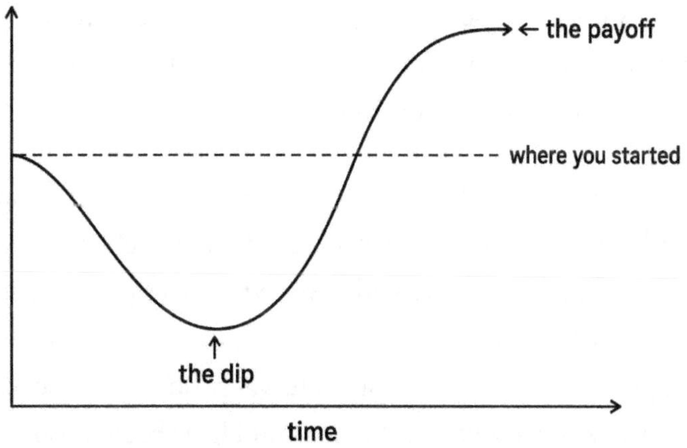

Figure 3.6 The J-curve, a dip before the payoff.

Telling a dip from a disaster

A real failure looks exactly like a dip at the start. Both go down. If you hold your nerve through everything, you'll ride a genuine failure all the way to the bottom, so you need a way to tell them apart, and it isn't the depth of the drop or how it feels. It's the shape over time. A dip has a bottom, it levels off and climbs back, and it recovers inside the window you expected when you worked out the delay. It has a reason you can name, and the reason is always learning. Your team, even while they struggle with it, wouldn't want to go back. A disaster is the opposite on every count. The number keeps falling past your window, you can't name a reason beyond the tool not doing what you needed, and people are already working around it to get their jobs done. When you see that pattern, stop calling it a dip, and stop funding it.

A few other signs tell you the same thing, that a project has stalled rather than dipped. The tool shows up in demonstrations but never in daily work. One person is the only one who really uses it. Checking its output takes as long as doing the task used to. The savings appeared in the first week and never came back. Any one of these can happen in a healthy rollout, but two or more together mean the same thing the falling number does, stop and take a hard look before you spend another dollar on it.

This is why it pays to make sharing failures normal, out loud and early, because a hidden failure is the expensive kind. We watched this happen with a business we advise. The owner had never written a single line of code in his life, but armed with an AI chat tool, he rebuilt the company's intake form himself and published it to the website. He was proud of the work, and he had every reason to be, since a year earlier it would have been impossible without hiring someone. What nobody realized for several days was that his change had overridden the styles and markup on several key landing pages, the ones running paid ads. The pages were still loaded, so nothing looked wrong from the outside, but they were now broken where they counted. The business kept paying for clicks that landed on pages that no longer converted, so the ad spend climbed while the leads coming back from it fell. Two kinds of damage at once, and neither showed up on the surface. One number would have caught it in a day, the leads coming in from those pages each week. A dip shows up in a count and recovers. Damage shows up in a count and keeps falling. Without the count, both look like a quiet week.

The Walk-Around Test

Before you open a single report, spend twenty minutes once a month walking the business and looking with your own eyes. This isn't measurement yet It's how you find what's worth measuring.

Ask yourself four things as you go. Where does work pile up now, and is it the same place it piled up three months ago? Who leaves on time this week who didn't used to? What gets checked twice, and does the checking now take longer than the doing? Which number in last month's books moved, and can anyone explain why?

Then ask the team one question you can't answer by watching. What did you stop doing that you're glad to be rid of? When adoption is going well, people name tasks that genuinely left. When it's going badly, they describe work that arrived instead, or they go quiet, and quiet is its own answer. If nobody can name a task that actually left, the work probably just moved to someone else, and it's worth finding out where.

The walk-around doesn't prove anything. It points you at the one or two places where something real is happening, so that when you sit down to measure, you're measuring the right thing.

Pick One Number

The walk-around tells you where to look. Now pick one number, from one task, and count it for a few weeks. The natural menu is small, hours spent on the task, errors caught, turnaround time, follow-up messages sent, leads coming in. Choose the one the walk-around pointed at, the place where something felt different.

Count it the same way every time, keep it in date order, and compare it against a baseline you wrote down before the change [17]. The date order matters more than it sounds. Two averages, one from before and one from after, hide everything useful. A line of weekly numbers shows you the shape, the dip, the recovery, and the week the improvement stalled. Those shapes are the difference between a project that's working and one that is going nowhere, and you saw earlier how much a single tracked number can be worth. The owner whose landing pages broke needed exactly one of these, the weekly count of leads, and it would have paid for itself a hundred times over in the first week alone.

The Three Doors

Time saved only becomes money by walking through one of three doors, and each door is a real line in your books you can point to.

A cost you stopped paying. Something that used to cost money now doesn't. Overtime dropped because the work gets done inside normal hours. A subscription you canceled because the AI tool replaced it. A contractor invoice that stopped arriving because the task came back in-house. If a real expense falls, you can find it on a statement, and that's banked money.

Capacity you sold. Your team can now handle more without growing, and you actually filled that room with paying work. More jobs completed per week. More clients served with the same headcount. The new customer you took on without hiring anyone. The test is that the freed-up capacity turned into an invoice. If it did, that's banked too.

A loss you avoided. Something that used to cost you money stopped happening. Fewer credits and refunds. Less warranty rework. The customer who was about to leave and stayed, was worth what they actually spent with you. These are harder to see because they're about things not happening, but they show up in your books as numbers that fell.

When Maria says show me the line, these are the lines. Any change that walked through one of these three doors is a real return, and you can name the line item that proves it. As Figure 3.7 shows, there's also a fourth door, one that isn't money yet.

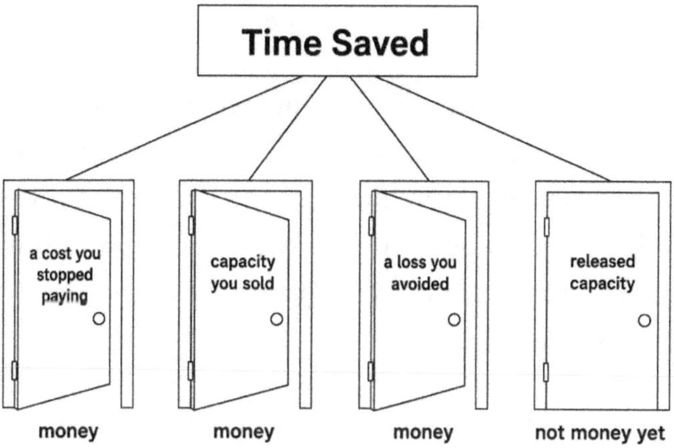

Figure 3.7. The four ways time saved can land in your business.

The Fourth Outcome

Here's the honest part most people skip. A lot of the time, saving someone time doesn't walk through any of the three doors. Payroll didn't fall, you didn't sell the freed hours, and no loss was avoided. The task just got easier.

That's still worth something, and it has a name: released capacity. It means you've freed up time or effort that hasn't yet turned into money. Your estimator gets two hours back every week, but she's still on the same salary, still working the same hours, just less stressed. Nothing changed in your books. What you have is room, capacity that exists but hasn't been sold, spent, or saved.

Released capacity is real and worth having. Easier days keep good people from leaving. Slack in the schedule lets you take on the next job without panic. Time freed in one place can be pointed at something that does make money later. The mistake is calling it return before it becomes one. Until that freed time walks through one of the three doors, turns into a cost you cut, work you sold, or a loss you dodged, it's capacity, not cash. Write it down honestly as what it is, and keep an eye on whether it ever converts.

This honesty protects you later. If you count released capacity as return, your year-one numbers look impressive and your year-two numbers can't back them up. When someone finally checks and the money isn't where you said it was, they don't just discount that one claim, they stop trusting the whole effort, and that's how a promising AI program loses its budget, not because it failed but because it was oversold.

Being straight about this also answers a question your team is asking whether they say it out loud or not, which is whether AI is here to replace them. The honest data is reassuring. In a US Census survey of businesses using AI, fielded over the winter of 2025 and 2026, sixty-six percent used it only to augment what workers do, and just two percent reported any AI-related drop in employment [18]. A freed-up hour can make someone's work better without being a headcount cut. Don't promise that no job will ever change, because you can't know that. The promise you can keep is that you'll tell people what you know, when you know it.

The Conversion Ledger

Everything in this chapter lands in one place, a single page you can put in front of anyone who asks what the AI has been worth. Call it the conversion ledger, because its job is to convert scattered impressions into lines a skeptic will believe.

One row for each change you've seen since the tools went in. Four columns across.

What you noticed. The change in plain words. Quotes go out faster. The inbox runs itself. Fewer callbacks.

What you counted. The number behind it, measured the same way over time. Quote turnaround fell from three days to four hours over six weeks. Not a feeling, a figure with a method.

Which door. How that number turns into money, if it has yet. A cost you stopped paying, capacity you sold, or a loss you avoided. If it hasn't walked through any of the three, the entry here is released capacity.

What it was worth. The dollar figure once it lands, pulled from a real line in your books. Or the honest words released capacity, if the money hasn't shown up. Figure 3.8 shows what a filled-in ledger looks like.

What you noticed	What you counted	Which door	What it was worth
quotes go out faster	3 days → same day	work sold	revenue +15%
fewer callbacks	callbacks −60%	cost cut	overtime −$18k/yr
returns dropped	returns −40%	loss avoided	savings +$67k/yr

Figure 3.8. Sample entries showing how each change moves across the ledger.

A row that reaches a dollar in the last column is return, and you can hand it to your accountant without flinching. A row that stops at released capacity is your honest answer about slack, easier days and a team with room to think, real and worth having, but not yet money. Both kinds of row belong on the page. A ledger that shows only wins reads like a sales pitch, and the point of this one is to be believed. The released-capacity rows are what make the banked rows credible.

Fill one row a week. In a few months the page tells you, in order, which projects turned into money, which are still promising, and which quietly stopped moving. That page is the answer to Maria's question, and it survives her pen.

Worksheet

The Conversion Ledger

A one-page format that ties each change to a number, a door, and a dollar.

Scan or visit **designated.ai/books/practical-strategy**

Six Months Later

At the next sit-down, half a year on, Maria started to ask the same question. Before she could finish, Luca moved the page right in front of her.

"Here you go."

Seven rows.

"Two of them are banked. The biggest one, we used AI to build and launch a new service line in about a quarter, work that would have taken us most of a year before. It's added roughly seventeen percent to revenue since it went live, and I can trace every dollar, because it's a line that didn't exist last time we spoke. The other five rows are released capacity. Real time freed up, easier weeks, but I'm not counting a dollar on them until they turn into something I can point to."

Maria read the page top to bottom. This time she had numbers she could actually write down. You could see it settle over both of them, a small nod of understanding. For the first time they were speaking the same language, the language of what the work was actually worth, and that was the thing that would serve the business from here on.

The number was real this time, not a feeling. Luca could defend every line of it, which is what six months of motion never gave him.

Measuring well gives you a short page of things AI is actually worth, in numbers you can stand behind, instead of a long list of things it is merely doing. It turns a feeling that the tools are helping into proof a skeptic will accept, and that proof is what earns the room to keep going. The more you lean on these tools, the more it matters that they run safely, and that speed does not turn into risk. That is what the next chapter is about.

Chapter 3.7

Build to Scale

Once you know how to measure the real value of your AI projects, the next challenge is making sure those early wins can grow into lasting improvements. In the following section, you will learn how to design your AI solutions so they can scale across your business, avoid common pitfalls, and deliver results that last.

Most AI pilots fail to scale because they're built as isolated solutions rather than as parts of a broader system. Teams often treat them as experiments designed to solve a single problem, without considering how they'll evolve or connect to other processes. Clear ownership is usually lacking, leaving no one accountable for maintaining or expanding the solution over time. Governance is often missing, too. There are no shared standards for how models are used, how data is managed, or how decisions are made. Even when a pilot demonstrates value, it often stalls because the organization isn't structured to support it beyond the initial phase.

FROM ISOLATED PILOT TO SCALABLE OPERATING MODEL

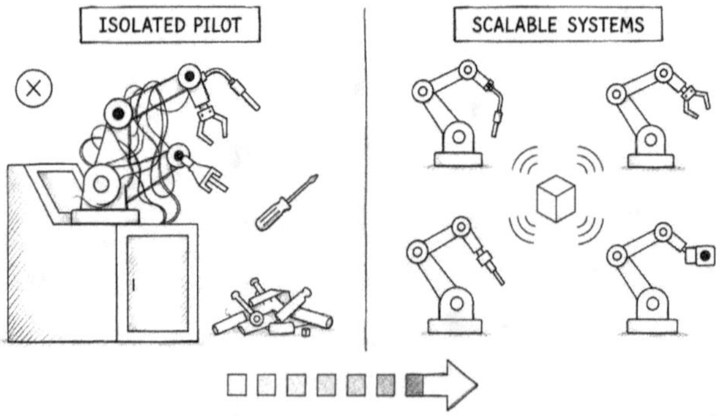

THE SHIFT TO DISTRIBUTED, REPEATABLE VALUE CREATION

Building AI to scale requires a different mindset from the start. Instead of creating one-off tools, organizations need to design systems that can be reused and adapted. This means separating models, data, workflows, and interfaces so each part can evolve without breaking the overall system. It also means planning for change, since models, vendors, and business needs will continue to shift. Beyond the technical side, scale depends on structure. Clear product ownership, defined decision rights, and simple processes for review, risk, and iteration are essential. The technical approach should remain modular and flexible, with the ability to swap tools or models as needed.

Scaling AI isn't about improving a single solution. It's about building the capability to repeatedly turn ideas into systems that deliver value. Organizations that succeed are those that treat AI as an operating model, not just a technology. As the saying goes, "You don't scale people, you scale systems."

Dimension	Scalable	Not scalable	Threats & risks	Cost of getting it wrong
Ownership	Named owner responsible for performance, updates, and decisions. Clear escalation path.	Built by one person who "just figured it out." No documentation, no handoff plan.	Key-person dependency. System degrades when builder leaves, gets promoted, or gets busy.	Full rebuild or abandonment. 2 to 6 months of lost momentum plus hiring/ retraining costs.

Architecture	Modular: data, logic, interface, and model are separable. Components can be swapped independently.	Monolithic: everything hardwired together. Changing one part risks breaking others.	Vendor lock-in. Cannot adapt when pricing changes, APIs deprecate, or business needs shift.	Forced migration under pressure. Typically 3 to 10x the cost of building modular from the start.
Data	Standardized inputs and outputs. Clean handoffs between systems. Data quality is monitored.	Works with one team's data format. Breaks when another team's data looks different.	Silent failures. Bad data produces bad outputs that go undetected, eroding trust and decisions.	Incorrect decisions made at scale. Cleanup costs grow exponentially the longer bad data flows.
Governance	Simple, written rules: who can change what, how changes are tested, how to roll back.	No rules. Anyone can modify. No testing process. No rollback plan.	Conflicting customizations across teams. Uncontrolled changes create unpredictable behavior.	Operational chaos at scale. Hours spent debugging issues that could have been prevented with a one-page protocol.
Portability	A second team can adopt it without the original builder present. Documentation exists.	Requires heavy hand-holding to replicate. Tribal knowledge, not documented process.	Scaling bottleneck. Growth is limited by the availability of a small number of people.	Linear cost growth instead of leverage. Each new deployment costs as much as the first.

Change readiness	Designed to evolve. Vendor, model, or business rule can change without full rebuild.	Brittle. Any change to one component requires rethinking the entire system.	Obsolescence. The tool or vendor landscape will shift within 12 months. Rigidity becomes liability.	Replacement cost equals or exceeds original build. Organization loses confidence in AI investments.
Value alignment	Tied to a measurable business outcome: time saved, decisions improved, risk reduced.	Scaled because it "worked" without defining what working means for the business.	Tool sprawl. Organization accumulates systems nobody fully understands or can justify.	Wasted budget and eroded leadership trust. Makes future AI investments harder to approve.
Team adoption	Teams trained, bought in, and supported. Feedback loops exist. Adoption is measured.	Rolled out by mandate. No training, no input, no support. "Just use it."	Shadow workarounds. Teams revert to old processes or build parallel systems that bypass the tool.	Investment with zero return. The system exists but nobody uses it. Political cost of the failed rollout lingers.

Scorecard

The Scale Readiness Scorecard

Rate your AI system across the seven dimensions that decide if it survives growth.

Scan or visit **designated.ai/books/practical-strategy**

Chapter 3.8
Guardrails That Matter

AI can help your business move faster, automate work, analyze information, and make better decisions. But as AI becomes more deeply connected to your operations, we face one critical question. What should AI be allowed to see, decide, and do?

This does not require a complex governance program. Most small businesses do not need heavy frameworks or long policies. What they do need are a few clear, practical rules that everyone understands and follows consistently.

Be Careful What You Give AI

Not all information is safe or appropriate to use with AI systems, and the key is understanding the difference between low-risk and high-risk data.

Public information such as marketing content, product descriptions, general research, and non-sensitive business material is usually safe to use. However, customer records, employee data, financial information, contracts, passwords, health records, confidential strategies, and proprietary company knowledge require much more caution because they can create serious privacy, legal, or competitive risks if exposed or misused.

Before entering sensitive information into any AI tool, understand how that tool handles data. Check whether the provider stores the data, whether it is used to train or improve their models, how long it is retained, who can access it, and whether administrators can control or delete it. Once data is shared with an external system, control over it is reduced.

If you would not feel comfortable sending the information to an outside company, do not paste it into an AI tool.

This becomes even more important when employees independently experiment with new AI tools. Even if a tool is useful, it can still introduce risk if no one has reviewed how it processes, stores, or protects company data. Without clear guidance, well-intentioned experimentation can unintentionally expose sensitive information.

Checklist

The Data Sensitivity Checklist

Sort what is safe to paste into AI from what is not, in one page.

Scan or visit **designated.ai/books/practical-strategy**

AI Can Be Wrong and Unfair

AI systems can produce answers that sound confident but are completely incorrect. They can also reflect or amplify biases present in the data they were trained on or the information they are given. This means AI outputs should always be treated as assistance, not absolute truth.

The risk becomes significantly more important when AI is used in decisions that affect people.

Using AI to summarize a meeting or draft a document is very different from using it to decide who should be hired, promoted, approved for a loan, flagged for fraud, or removed as a customer. In these cases, the consequences of mistakes or bias are much higher, and the need for oversight increases.

The higher the impact of a decision, the more important human involvement becomes. AI can support decision-making, but it should not replace human judgment in high-stakes situations.

Employees should know when they must verify AI outputs, when a manager must review or approve results, and when AI is only allowed to provide recommendations rather than final decisions. Without this clarity, there is a risk of over-trusting AI or assuming it is more accurate than it actually is.

Transparency is also important. If customers or employees are interacting with an AI system, they should be informed in appropriate situations. Businesses should avoid creating situations where people believe a human made a decision or review when the process was primarily driven by AI. Clear communication builds trust and reduces misunderstanding.

Control What AI Can Do

AI is evolving from simple assistants that generate text into more powerful systems that can take actions. These AI agents can send emails, update databases, access customer records, create documents, process transactions, schedule meetings, and trigger automated workflows.

An AI assistant that drafts an email is relatively low risk because a human can review the message before it is sent. An AI agent that can send emails directly, modify customer records, issue refunds, or approve financial transactions introduces a much higher level of responsibility and potential impact.

AI should always be given the minimum level of access required to perform its function. It should not have broad or unrestricted permissions unless absolutely necessary.

Some actions can be automated, while others should require human approval before execution. High-impact actions involving money, customers, employees, legal commitments, security, or company reputation should always include clear checkpoints and oversight.

AI access should follow the same principle as employee access. Not everyone in a company can see or change everything, and AI should not either. Access should be role-based, limited, and intentional.

Create Simple Rules People Can Follow

AI governance does not need to start with complex documentation or formal frameworks. Clarity and consistency matter more than length or complexity.

Protect sensitive data. Confidential, personal, financial, or regulated information should not be entered into unapproved AI tools.

Verify important outputs. AI-generated facts, numbers, recommendations, and customer-facing content should be checked when errors could lead to meaningful consequences. Responsibility for accuracy remains with people.

Keep humans involved in high-impact decisions. AI can support decisions involving people, money, safety, legal matters, or reputation, but it should not be the final decision-maker when consequences are significant.

Limit access. AI systems and agents should only be given the data, tools, and permissions they need to perform their specific tasks. Excess access increases risk without adding value.

Assign responsibility. Every important AI system should have a clearly identified owner who understands how it is used, monitors its behavior, and knows how to respond if something goes wrong. Without ownership, issues can go unnoticed or unresolved.

The goal of these rules is not to slow down innovation or eliminate risk entirely. The goal is to understand where the risks are meaningful and put appropriate protections in place.

When done correctly, guardrails do not restrict AI adoption. They create confidence. And with confidence, organizations can move faster, experiment more safely, and scale AI use responsibly.

Template

The Simple AI Rules Template

Six one-line rules your team can follow starting Monday.

Scan or visit **designated.ai/books/practical-strategy**

Chapter 3.9
What Comes Next

Looking into the future is always exciting, especially after experiencing this much innovation in just five years. Some technologies remain largely untapped. Quantum computing is the clearest example, and any breakthrough there opens completely new horizons, from drug discovery to data security. The pace we have seen so far is the slowest it will ever be.

At present, businesses face an overwhelming number of AI tools, far more than anyone can evaluate, let alone adopt. More people than ever are being educated in these technologies, yet many still struggle to fully grasp the underlying concepts. That gap closes from the tool side. Technical complexity moves out of sight, powerful tools get simpler to use, and working solutions come faster. The number of companies offering trusted tools consolidates, since building effective solutions no longer requires deep engineering effort.

The models underneath evolve too, into more advanced systems that increasingly train themselves, whatever names they carry by then. They rely not only on curated datasets but gather information autonomously from a wide range of sources. This may feel unsettling, but it is a natural progression of the technology and, to a large extent, unavoidable. As these systems grow more powerful, guardrails matter more, especially given the cultural norms and ethical boundaries that took centuries to develop.

Our Friendly Bets

Throughout this book we took an evergreen approach, focusing on principles and patterns built to stay relevant for the next seven to ten years rather than tools and trends with a short shelf life. Predicting exactly what happens is much harder. Anything we state with confidence will read as either obvious or foolish a few years from now.

Instead of pretending otherwise, we decided to have some fun. Each of us places a few friendly bets on what the world looks like in five to ten years, staked in the territory he knows best. Mock us in 2033 if we're wrong. That's part of the deal.

Marcin

Sensors and robots run the physical world.

Think about the last time you circled a mall lot on a Saturday. By 2032 the first hotels and hospitals run garages where cars park themselves, guided by cameras and sensors, fitting tight spaces no one could use when doors still had to open. Airports, offices, and eventually your mall follow. You tap your phone from the food court and the car meets you at the entrance. Valet parking is no longer a job, and every function built on a person watching, checking, or collecting goes the same way.

Robots take over the dangerous work first, fighting fires, inspecting collapsed structures, handling hazardous materials, and then show up in ordinary places, greeting you at reception desks and restocking shelves overnight.

Software consolidates into agents.

Today Anthropic, OpenAI, and Google ship frontier models that generate work previously reserved for specialized software, designs, presentations, spreadsheets, code, full documents. That trend continues. By 2031 your company works inside one AI system that covers what 20 to 40 apps handle today, connected to everything your business knows. Need a simple narrow app? You code it yourself, no programming experience required. Basic SaaS goes first. Complex systems survive because of the data, the security, and the orchestration woven through the whole business. Your list of subscriptions shrinks, and your AI credits bill goes significantly up.

We carry digital twins with us.

Today, when your best person leaves, everything she knows leaves with her. By 2031 that stops. Companies keep a working memory of how things get done, and people build digital twins of their own expertise that travel with them from job to job, growing the way a reputation grows. Both sides win. The business keeps what it learned, and your experience becomes something you own for life. Somewhere along the way, the first team starts reporting to an AI boss that runs the schedule and pulls in a human only when judgment is needed. I'm betting my next company on this one.

Medicine merges with the machine.

Over 200 AI-discovered drugs sit in clinical trials as of mid-2026, and the first brain implants are already restoring movement and speech to paralyzed patients. Both trends accelerate from here. By 2033 AI helps deliver a cure for a major disease that has resisted medicine for decades. Chips in human brains move from medical necessity to elective choice, and the first person pays to connect a healthy mind to a machine because of what it adds. The line between using AI and being connected to it starts to blur.

Tarik

Today, we still chat with an AI, type a prompt, wait, and get an answer. In the near future, this back-and-forth will start to disappear. We will spend less time prompting AI and more time delegating to it. We will give an AI agent a clear goal, and it will figure out the steps, choose the right tools, manage the tasks, and do the heavy work for us. We will move from asking AI to help us do the work to deciding what work needs to be done.

This change will give small businesses a huge advantage. Small teams will be able to do things that used to require large departments, big budgets, and specialized teams. A small company can use AI for marketing, research, data analysis, customer support, and operations at a high level. We will see small, agile companies compete directly with much larger corporations.

For organizations, adapting will not be a choice. The pressure will come from customers who want faster and more personalized service. Smaller competitors will move faster and operate at lower costs. Employees will expect better tools, and governments will introduce new rules. Many companies will not transform because they want to innovate. They will transform because the world around them forces them to.

At the same time, as automation grows, human authenticity will become more valuable. We will appreciate a meal cooked by a chef more when machines can make thousands of meals. We will value live music more when platforms are full of AI-generated songs. Real craft, genuine conversations, and human experiences will not disappear. They may become more valuable simply because they are human.

Finally, we cannot assume that every technological advancement will create a better world. The same tools that make businesses faster and more capable can also cause harm when they are used poorly. We need good rules, responsible leadership, and systems that protect people while still allowing innovation to grow. Technology will keep improving quickly. Our job is to make sure we use it wisely.

Jens

If we look five years into the future, many of the innovations we currently see in scientific papers will begin to enter everyday reality.

One major step will likely be the introduction of well-trained humanoid robots to the public. These systems could assist in households, gardening, security, shopping, and many other practical applications. For those who enjoy small pets, there may even be tailored robotic solutions to support their care. Personally, I can easily imagine owning one for gardening, that's probably the task I find most tedious.

At the same time, progress in artificial intelligence will continue at an extraordinary pace. Since AI development depends primarily on computational resources rather than specialized hardware, research communities worldwide are constantly improving model architectures and training methods. As a result, innovation cycles will become significantly shorter than what we are used to today. Looking slightly beyond what seems immediately achievable, humanoid robots could evolve into affordable and reliable workers, particularly for repetitive tasks across various industries.

Extending the horizon a bit further, quantum computing may begin to play a transformative role. By drastically reducing computation times, it could enable the rapid development and training of highly complex models. In medical research, for example, efforts are already underway to detect diseases that currently evade standard diagnostic methods.

The idea that a single drop of blood could provide a comprehensive health analysis is no longer purely speculative, it is actively being explored.

Quantum technologies may also revolutionize secure communication, with concepts like quantum entanglement enabling fundamentally new levels of data security. At the same time, the cycle for identifying and responding to new diseases could be dramatically shortened.

Overall, the outlook is highly promising, suggesting a future defined by rapid innovation, increased efficiency, and entirely new technological capabilities.

Chapter 3.10

Continue the Journey

Not long ago, I was sitting in a classroom at MIT Professional Education pursuing my Chief Product Officer program. The program had drawn executives from around the world, including senior leaders from some of the largest technology companies on the planet. People whose companies shape how the rest of us interact with it. And every single time a session turned to artificial intelligence, something shifted in the room. Laptops opened. Phones went down. Conversations stopped. Everyone leaned in.

It struck me then, and it has stayed with me since: nobody has figured this out. Not the largest companies. Not the most credentialed leaders. Not us. Everyone is learning as they go, building as they discover, adjusting as they understand more.

Be Curious

The difference between the leaders who move forward and the ones who stall is not knowledge. It is a curiosity. The willingness to ask a question without already knowing the answer. The willingness to try something without a guarantee that it will work. The willingness to look at a tool, a workflow, or a business problem and wonder: what if we approached this differently?

You do not need a technical background to be curious. You do not need a certification. You need the same instinct that made you start or lead a business in the first place: the belief that you can figure things out as you go, as long as you keep going.

If the executives at the world's largest technology companies are still leaning in to learn, you have every right to be in the same room. You already are.

Start Today

Do not start with a course. Do not start with a strategy. Start with a question.

Open one of the AI assistants available to you today. Claude, ChatGPT, or Gemini. All three are free to start. Bring a real business question. Not a test prompt. Something you are actually dealing with this week. A hiring decision. Customer communication. A process that feels heavier than it should. Ask the tool to help you think through it.

You will learn more about what AI can and cannot do in 30 minutes of hands-on use than in hours of reading about it. The point is not to get a perfect answer. The point is to start building an intuition for where these tools fit inside your work. That intuition only comes from doing.

Build, Break, Adjust

We build. We find what works. We pivot when it does not. We change direction when the evidence indicates. That is not uncertainty. That is how real progress has always happened.

Every business leader in this space is writing the playbook as they go. The ones who succeed are not the ones who wait for the perfect plan. They are the ones who tried something, learned from it, adjusted, and tried again. The cost of waiting for certainty is higher than the cost of starting imperfectly.

Do not be intimidated by what you have not tried yet. Be encouraged by what you now understand well enough to try.

Join the Journey

This book was our starting point, too. The three of us came from different backgrounds, different industries, and different corners of the world. What brought us together was a shared conviction that AI adoption should not be reserved for companies with unlimited budgets and dedicated engineering teams. The businesses that power communities, employ neighbors, and serve real people deserve the same clarity and the same access to what works.

Writing this book was how we started sharing what we have learned. But it is not where we stop. Through Designated AI, we are building a community for the people who are actually doing this work. Not watching from the sidelines. Not waiting for someone else to go first. Doing it.

We are building this for three kinds of people:

Practitioners who are implementing AI inside their businesses and want to learn from others doing the same.

Providers who help businesses adopt AI and want to stay connected to what real operators actually need.

Mentors who have been through the journey and are willing to share what they learned with those just getting started.

We invite you to be part of it. Not as a spectator, but as a contributor. The best ideas in this book did not come from research alone. They came from doing the work, getting it wrong, adjusting, and trying again. The next set of best ideas will come from you.

There is a quote, often attributed to Gandhi, that has guided how we approached this project and how we think about the work ahead:

"Live as if you were to die tomorrow. Learn as if you were to live forever."

Act with urgency. Learn without limits. The journey does not end here. It begins.

Webpage

Designated.AI - Resource Center

Scan or visit **designated.ai/books/practical-strategy**

References:

1. Rogers EM. Diffusion of Innovations. 5th ed. Simon & Schuster; 2003.

2. Goldsmith RE, Hofacker CF. Measuring consumer innovativeness. J Acad Mark Sci. 1991 Jun;19(3):209–21. doi:10.1007/BF02726497

3. Jussupow E, Spohrer K, Heinzl A. Identity Threats as a Reason for Resistance to Artificial Intelligence: Survey Study With Medical Students and Professionals. JMIR Form Res. 2022 Mar 23;6(3):e28750. doi:10.2196/28750

4. SimanTov-Nachlieli I. More to Lose: The Adverse Effect of High Performance Ranking on Employees' Preimplementation Attitudes Toward the Integration of Powerful AI Aids. Organ Sci. 2025 Jan;36(1):1–20. doi:10.1287/orsc.2023.17515

5. Edmondson A. Psychological Safety and Learning Behavior in Work Teams. Adm Sci Q. 1999 Jun;44(2):350–83. doi:10.2307/2666999

6. IEEE Standard Model Process for Addressing Ethical Concerns during System Design [Internet]. IEEE; [cited 2026 Aug 10]. Available from: https://ieeexplore.ieee.org/document/9536679/ doi:10.1109/IEEESTD.2021.9536679

7. The seven operating truths of AI-native companies | McKinsey [Internet]. [cited 2026 Aug 17]. Available from: https://www.mckinsey.com/capabilities/business-building/our-insights/the-seven-operating-truths-of-ai-native-companies

8. Sadovski E, Aviv I, Hadar I. Navigating the Human-oversight Dilemma in AI-based Systems. In: 2025 IEEE 33rd International Requirements Engineering Conference Workshops (REW) [Internet]. Valencia, Spain: IEEE; 2025 [cited 2026 Aug 10]. p. 454–61. Available from: https://ieeexplore.ieee.org/document/11190219/ doi:10.1109/REW66121.2025.00069

9. Raiaan MAK, Mukta MdSH, Fatema K, Fahad NM, Sakib S, Mim MMJ, et al. A Review on Large Language Models: Architectures, Applications, Taxonomies, Open Issues and Challenges. IEEE Access. 2024;12:26839–74. doi:10.1109/ACCESS.2024.3365742

10. Chamola V, Hassija V, Sulthana AR, Ghosh D, Dhingra D, Sikdar B. A Review of Trustworthy and Explainable Artificial Intelligence (XAI). IEEE Access. 2023;11:78994–9015. doi:10.1109/ACCESS.2023.3294569

11. Malin B, Kalganova T, Boulgouris N. A Review of Faithfulness Metrics for Hallucination Assessment in Large Language Models. IEEE J Sel Top Signal Process. 2025 Oct;19(7):1362–75. doi:10.1109/JSTSP.2025.3579203

12. Humlum A, Vestergaard E. Still Waters, Rapid Currents: Early Labor Market Transformation under Generative AI [Internet]. Cambridge, MA: National Bureau of Economic Research; 2025 May [cited 2026 Aug 17]. p. w33777. Report no.: w33777. Available from: http://www.nber.org/papers/w33777.pdf doi:10.3386/w33777

13. Banker RD, Potter G, Srinivasan D. Association of Nonfinancial Performance Measures with the Financial Performance of a Lodging Chain. Cornell Hotel Restaur Adm Q. 2005 Nov;46(4):394–412. doi:10.1177/0010880405275597

14. Ittner CD, Larcker DF, Meyer MW. Subjectivity and the Weighting of Performance Measures: Evidence from a Balanced Scorecard. Account Rev. 2003 Jul 1;78(3):725–58. doi:10.2308/accr.2003.78.3.725

15. Bureau UC. Census.gov [Internet]. [cited 2026 Aug 17]. The Rise of Industrial AI in America: Microfoundations of the Productivity J-curve(s). Available from: https://www.census.gov/library/working-papers/2025/adrm/CES-WP-25-27.html

16. Brynjolfsson E, Rock D, Syverson C. The Productivity J-Curve: How Intangibles Complement General Purpose Technologies. Am Econ J Macroecon. 2021 Jan 1;13(1):333–72. doi:10.1257/mac.20180386

17. Perla RJ, Provost LP, Murray SK. The run chart: a simple analytical tool for learning from variation in healthcare processes. BMJ Qual Saf. 2011 Jan;20(1):46–51. doi:10.1136/bmjqs.2009.037895

18. Bureau UC. Census.gov [Internet]. [cited 2026 Aug 17]. The Microstructure of AI Diffusion: Evidence from Firms, Business Functions, and Worker Tasks. Available from: https://www.census.gov/library/working-papers/2026/adrm/CES-WP-26-25.html